CANNABIS
COOKBOOK

Delicious Edible Medical Marijuana Recipes for Beginners

By Tom Gordon

Table of Contents

Introduction viii

Chapter 1 - Benefits and Side Effects of Cannabis...........1

Benefits of Cannabis..1
Side Effects 2

Chapter 2 - Cannabis Herbs & Dispensary 4

Cannabis Sativa..4
Cannabis Indica 4
Cannabis Ruderalis...5
Marijuana Dispensaries 5
While Buying Cannabis 5
Choosing the Right Dosage...6

Chapter 3 - Basic Cannabis Recipes...................................7

How to Decarb Cannabis 7
Cannabis Tincture with Grain Alcohol.......................................8
Cannabis Tincture with Coconut Oil..9
Canna-Oil (Cannabis Infused Cooking Oil)..............................10
CBD Oil 11
Canna-Butter...13
Canna-Honey...14
Weed Sugar 15
Canna-Flour..16
Canna Milk (Cannabis Infused Milk) 17
Canna-Cream Cheese 18
Cannabis Mayonnaise 19
Cannabis Peanut Butter...20
Cannabis-Infused Apple Cider Vinegar.....................................21
Marijuana Vinaigrette 22
Italian Cannabis Dressing...23
Simple Salad Dressing 24
Sweet Salad Dressing...25
Honey Mustard Salad Dressing...26
Cannabis-Infused BBQ Sauce...27
Hazelnut Spread 28
Cannabis Salsa 29
Cannabis Caramel Sauce 30
Cannabis Glycerin Concentrate 31

Chapter 4 - Breakfast Recipes..32

Weed Crepes...32
Cannabis Waffles..34
Weed French Toast..36
Weed Pancakes...38
Cannabis Granola Breakfast Bars..40
Blueberry Muffins with Weed Streusel...............................42
Weed Breakfast Casserole...45
Breakfast "Baked" Burritos...47
Veggie Frittata...48
Kale and Chevre Goat Cheese Frittata...............................50
Weed Omelet..52
Banana Nut Bread...54
Berry and Banana Smoothie..56
Green Juice...58
Sweet n Spicy Juice...59
Strawberry Banana Smoothie..60
Raw Cannabis Green Smoothie...61
Weed Iced Coffee..62
Indian Bhang...63

Chapter 5 - Dinner Recipes 64

Cannabis-Infused Butter Chicken 64
Cannabis Chicken Fajitas 67
Easy Cannabis-Infused Chicken Adobo.............................70
Chicken Pot-Cacciatore..72
Cannabis-Infused Thai Green Curry...................................74
Classic Cannabis Lasagna...77
Cannabis-Infused Turkey Bolognese 79
Beef and Bean Marijuana Chili...81
Canna-Butter Pan-Seared Steak 83
Canna-Burgers...84
Weed-Infused Pulled Pork...86
BBQ Pork Ribs..88
Cannabis Garlic and Rosemary Pork Chops......................89
Jambalaya..90
Baked Shrimp Scampi...92
Grilled Fish Tacos with Ganja Green Salsa 94
Weed Fish and Chips...96
Sativa Shrimp Creole 98
Butter Garlic Shrimp 100

Cannabis-Infused Pasta with Clams and Green Chiles 102
Weed Grilled Cheese...105
Smoked Mac 'n' Cheese..106
Marijuana Pizzadillas..108
Cannabis-Infused French Bread Pizza..109
Fettuccine Alfredo Pasta...111
Mini Cannabis Green Bean Casserole..112
Mashed Cannabis Cauliflower with Parmesan Cheese......................114
Cannabis Mashed Potatoes...115
Caramelized Brussels Sprouts 116
Canna-Butter Sautéed Mushrooms 117
Broccoli Cheddar Cannabis Casserole 118
Cannabis-Infused Radical Ratatouille 120
Almond, Orange, and Cucumber Stuffed Avocado 122
Weed Bread..123
Cannabis Gravy 125
Weed French Fries 126

Chapter 6 - Dessert Recipes..128

Chocolate Pudding..128
French Toast Cupcakes..130
Key Lime Kickers 132
Brownies...134
Canna Chocolate Dipped Strawberries..136
Pineapple Upside-Down Cake 137
Macadamia & White Chocolate Cookies 139
Marijuana Chocolate Chip Cookies 141
Weed Chocolate Bars 143
Vegan Pumpkin Spice Ice Cream 144
Weed Vanilla Ice Cream..145
Banana Marijuana Ice Cream 147
Hash Fudge..149
No-Bake Cannabis Pumpkin Pie 151
Apple Pie 153
Cannabis-Infused Chocolate Cake 155

Chapter 7 - Salad & Soup Recipes...157

Cannabis Chicken Salad 157
Thai Mango Salad..159
Green Leafy Kale Salad with Brown Canna-Butter Vinaigrette...........160
Caesar Salad 161

Weed Salad...163
Strawberry Burrata Salad...165
Spinach and Orange Salad with Grilled Salmon and Orange Vinaigrette167
Cranberry Walnut Salad 169
Canna-Quinoa Salad...171
West Coast Garden Salad with Cannabis Olive Oil & Dill Dressing..173
Almond Apricot Chicken Salad 175
CBD Salad...177
Multiple Sclerosis (MS) Arugula Goat-Cheese Salad with Cannabis Citrus
Vinaigrette...179
Infused Fruit Salad..181
THC Tuna Salad..182
Marijuana Tortilla Soup with Vegetables.....................................183
Marijuana French Onion Soup Au Gratin 185
Cannabis Chicken Noodle Soup...188
Greek Lemon Chicken Soup..190
Vegan Split Pea Soup...192
Tomato Soup with Carrots and Celery 194
Hearty Vegan Winter Vegetable Soup 196
Cream of Cannabis Soup 198
Cannabis Tomato Soup..200
Green Cannabis Soup...202
Weed Ramen Noodles..204
Vegan Creamy Cannabis-Infused Potato Soup 205
Cannabis Mushroom Soup 206
Cannabis-Infused Bone Broth 208
Butternut Squash Soup with CBD Drizzle 210

Chapter 8 - Snack Recipes **212**

Weed Deviled Eggs...212
Weez-Its..214
Jalapeno Pot Poppers 215
Marijuana-Infused, 3 Layered Popsicle..217
Watermelon and Lime Cannabis-Infused Popsicles 219
Baked Apricot Brie 220
Cannabis-Infused Salted Caramel Popcorn..................................222
Monster Munchie Balls 224
Cinnamon Muffins...225
Weed Mozzarella Sticks...227
Weed Potato Chips 229
Weed Biscuits...230

Super Lemon Haze Mexican Guacamole 232
Pepper Crackers 233
Chocolate Crackers 235
Lemon Crackers 237
Sweet, Spicy, and Sativa Mixed Nuts 239
Cannabis Granola Bars..240
Hash Yogurt...243
Hush Puppies...244
Marijuana Hot Wings 246
Peanut Butter Balls 247

Conclusion.. **249**

INTRODUCTION

In recent times, there has been an increasing push all around the world for the legalization of Cannabis. It is already legal in countries like Canada, Jamaica, Uruguay, and many states in the United States.

We know you might have a lot of questions about Cannabis. Is it only used for getting high? Why are countries slowly legalizing Cannabis? Can it be helpful in relieving health problems? Well, we are here to answer these questions for you.

Cannabis is safe for medical use. In fact, it is known to be very helpful in relieving health problems and reducing pain. The medicinal value is far more significant than the harmful effects. Cannabis is quite known for its medicinal properties, and many doctors around the world recommend the intake of Cannabis in small amounts, as it can be quite helpful treating the symptoms of several medical ailments. It has helped reduce the suffering of thousands of people suffering from deadly diseases such as cancer, AIDS, and many more medical conditions. It is important to note that while Cannabis will not get rid of the disease, it will help to reduce the suffering. Apart from the medicinal benefits, economists are also of the opinion that legalizing Cannabis has the potential for turning into a billion-dollar industry in a very short while.

The cannabis plant is also called marijuana or Indian hemp. The botanical name is Cannabis Sativa / Cannabis Indica. More than 400 natural compounds are found in the cannabis plant. Tetrahydrocannabinol THC and Cannabidiol (CBD) are a couple of compounds that are generally used medically.

Now, while Cannabis has several benefits, it also has a downside. It has a very intense bitter flavor. So, how does one ingest Cannabis in a way that doesn't require smoking or ingesting it directly, but a way that's actually enjoyable and pleasing to the senses?

Through food, of course! Yes, Cannabis can be incorporated into your breakfast, meals, snacks, and even desserts. How? That's what we're here to tell you. This book contains several recipes that incorporate Cannabis into delicious foods that you can enjoy eating without the bad taste and, at the same time, also receive the medical benefits.

Read on to find out more!

CHAPTER 1

BENEFITS AND SIDE EFFECTS OF CANNABIS

Before we begin, let us first understand everything you need to about cannabis, starting with its benefits.

Benefits of Cannabis

Stimulates Appetite

Cancer patients, who undergo chemotherapy generally have a poor appetite and are not able to eat well. Cannabis helps them restore their appetite.

Reduces Pain

Cannabis can greatly help ease pain in patients who suffer from cancer, muscle spasms, chronic pain, terminally ill patients, etc. It also helps fight certain cancers like lung cancer and brain cancer.

Glaucoma

Helps treat glaucoma.

Parkinson's Disease

Helps reduce tremors.

Insomnia

It helps induce relaxation and sleep, especially in terminally ill patients.

Reduces anxiety and depression.

Alzheimer's Disease

Slows down the progression of Alzheimer's disease and can improve motor skills.

Bone Problems/Joint Pain

Helps with problems like arthritis pain and broken bones.

Diabetes

It can help prevent diabetes. It also helps control diabetes if you are a diabetic.

Nausea and Vomiting

Helps control nausea and vomiting.

Seizure Disorders Like Epilepsy

Helps to reduce seizures.

Autism, ADHD, ADD

Has a calming effect on patients with autism, ADHD, and ADD, especially when they are violent, acting out, or aggressive. Helps them to concentrate and focus better.

Side Effects

Long-term usage of cannabis can hinder the development of the brain. Memory problems and learning disorders are common in growing children. There is a tendency for schizophrenia in adolescents. Sometimes consuming cannabis can have an effect on medications you are taking. It is especially common for those adults who consume blood thinners.

There is always a chance of overdosing. When you consume edible cannabis, it takes longer to start showing the effect on your body (takes a couple of hours) than when you smoke it (it takes only a few minutes). So people tend to overdose when they do not find immediate relief. Overdosing can cause nausea, psychosis, panic attacks, hallucinations, etc.

Long-term ingestion of cannabis can cause irregular heartbeat, heart attacks, increased heart rate, low blood pressure, high blood pressure, etc. Your reflexes can become slower. You may feel less energetic and tired all the time.

A couple of weeks of cannabis use can make you dependent on it, which might further lead to addiction. Suddenly stopping the intake of Cannabis can cause severe withdrawal symptoms.

Consuming cannabis in bipolar disorder patients can worsen their disorder.

Immunity can be weaken.

Mental health may get affected. If you already have mental health problems, it can worsen. You may not be able to think or judge properly. You may suffer from memory loss, schizophrenia, hallucinations, or paranoia.

Eyes can become red or dry. You may have blurred vision. You may feel drowsy, fatigued, or dizzy.

You may have a sore throat or cough. Your taste buds may be affected.

Diarrhea or constipation is common. Overdosing can cause nausea and vomiting.

Some patients may have a skin rash.

Using cannabis with other medicines may have side effects.

CHAPTER 2

CANNABIS HERBS & DISPENSARY

The different types of cannabis are:

1. Cannabis Sativa

2. Cannabis Indica

3. Cannabis Ruderalis

There are also male plants and female plants. There isn't much difference in the species of cannabis. The species are more or less the same. It is more about where it is grown because of climatic conditions; the plants have different height.

Cannabis Sativa

They are tall plants (about 10 - 15 feet in height) with lengthy, thin leaves and soft seeds. This variety goes well with cooking. It can be smoked as well. One generally feels happy and energetic after consuming it. It is high in THC. This variety of cannabis helps in overcoming depression and mood disorders. It is also good for those with ADHD. It is cultivated in equatorial regions like India, Mexico, Thailand, etc. The flowers take time to grow.

Cannabis Indica

These plants do not grow as tall as Cannabis Sativa. They grow in bushes, and no is more than 6 feet in height. They have short, rounded leaves and soft seeds. These plants can be grown indoors. The flowers do not take as much time to grow as Cannabis Sativa. These are cultivated in Asian countries that have short winters like Nepal, Afghanistan, and Lebanon and also in North African countries like Morocco. The buds and flowers grow in clusters and are sticky compared to Cannabis Sativa. Because of its healing properties, it is generally recommended. It is of great help in pain relief. It has a calming effect and great for curbing anxiety and putting you to sleep.

Cannabis Ruderalis

These plants are even shorter than Cannabis Indica. The height of the plants is about 1 ½ - 2 feet. The flowering happens earlier (within 3 – 4 weeks of sowing the seeds) compared to Cannabis Indica. They grow naturally in Russia and are considered weeds. They are also cultivated in Central Europe, Eastern Europe, and Central Asia. This variety of cannabis is great for calming and relaxing.

In spite of having these differences, they all can be grown together. Hybrid varieties of cannabis are also being cultivated nowadays. In fact, hybrid cannabis is more commonly and popularly grown as compared to their individual 3 varieties.

Marijuana Dispensaries

When you need to buy cannabis, buy it from a state where it is legal. Make sure it is from a licensed dispensary. Buy the kind that suits your personal needs. If you do not buy the right form, it can affect your health. You need a doctor's prescription for consuming cannabis, so don't forget to carry any other documents that are required in the state that has legalized it. Registration at the dispensary is necessary.

While Buying Cannabis

Keep the following aspects in mind before buying cannabis.

- Buy only good quality cannabis. It should look good and have a fresh smell, but not all dispensaries allow you to touch or smell it. Do not buy anything that looks old or has an awful smell. Good quality cannabis is expensive.

- If possible, visit a dispensary to make your purchase rather than ordering home delivery.

- Before going to a dispensary or finalizing your dispensary, check the reviews on the Internet. If you find many negative reviews, look for another dispensary.

- When you visit the dispensary for the first time, find out where they procure the cannabis if they do not grow it themselves. Find out the types of cannabis they have.

- You need to be at least 18 years of age to visit the dispensary.

- Since you are using it for medical purposes, you need not pay taxes.

- Qualified budtenders at the dispensaries can help you choose the right cannabis.

Choosing the Right Dosage

Consuming the same quantity of cannabis can have a different effect on different people. The effect varies from person to person, whether men or women. So you cannot have a generalized dose. The same goes for the withdrawal symptoms as well. It can vary from person to person. Dosage also depends on how frequently and for what purpose you use it.

Your medicinal practitioner will not generally prescribe the exact dosage. He will generally recommend that you start with low doses, taking into consideration your age, medical conditions, the other medications you are on, etc. The medicinal practitioner will also consider if cannabis will have adverse effects on your current medicines. Ingesting very small doses is called micro-dosing, which is may be equivalent to 1/20 of a normal dose. You need to wait for 4-5 hours before taking the next dose. To start, take no more than 3 doses in 24 hours. Make sure to drink lots of fluids. In a few days, you will know how it is affecting your body and whether it is helping your problem and how well you are able to tolerate it. You can now decide if you want to lower the dose or increase the dose. This is called dose titration.

In most patients, low doses work well, but higher doses can have an adverse effect. Measure the cannabis tincture (when mentioned in any of the recipes) using a pharmacy dropper.

CHAPTER 3

BASIC CANNABIS RECIPES

Always consult your medical professional before consuming cannabis.

Always adjust the cannabis in the recipes to suit your personal needs.

How to Decarb Cannabis

Preparation time: 2 minutes

Cooking time: 35 – 60 minutes

Makes: As required

Ingredients:

- Cannabis (marijuana), as required

Directions:

1. To prepare oven and baking sheet: Place a sheet of parchment paper on a baking sheet. Set up your oven and adjust the temperature to 220° F. Let the oven preheat.

2. Grind the cannabis to a fine powder.

3. Spread powdered cannabis all over the baking sheet. Place the baking sheet in the oven and bake for 35-60 minutes. You can bake it for longer if desired.

4. Remove from the oven. This is what you call decarboxylated or decarbed marijuana/ cannabis. Follow this procedure when decarbed is mentioned in the recipe.

Cannabis Tincture with Grain Alcohol

Preparation time: 5 minutes

Cooking time: 0 minutes

Makes: about 1 ½ cups

Ingredients:

- ½ ounce cannabis, ground into smaller pieces, decarbed

- ½ quart grain alcohol like Everclear

Directions:

1. Sterilize a quart size glass jar or mason's jar.

2. Add cannabis into the jar.

3. Pour alcohol into the jar.

4. Tighten the lid and place it in a cool, dark area like in a cupboard for about 2 – 3 weeks. You need to shake the jar every 2 – 3 days.

5. Strain with a cheesecloth or a fine wire mesh strainer into a beaker or measuring cup with a spout.

6. Pour the tincture into dropper bottles. Place in the refrigerator or a cool place until use.

Cannabis Tincture with Coconut Oil

Preparation time: 5 minutes

Cooking time: 6 – 8 hours

Makes: About 1 ½ cups

Ingredients:

- 2 cups cannabis, ground into smaller pieces

- 2 cups coconut oil

Directions:

1. To set up double boiler: Take 2 pots of nearly the same (but not same) sizes such that the smaller one fits inside, the larger pot.; the smaller pot should not touch the bottom of the bigger pot. It should fit well inside it.

2. Pour enough water into the larger pot such that it is 1/3 full. The water should not touch the smaller pot. Place the bigger bowl over medium flame. Let the water come to a boil.

3. Add coconut oil into the smaller pot. Place the smaller pot inside the bigger pot.

4. Lower heat to low heat and let the water simmer. Let it simmer this way for 6 – 7 hours. This is another method for decarboxylation.

5. Remove the inner pot from the double boiler and let it cool.

6. Line the top of the storage container with a piece of cheesecloth. Pour oil into it and discard the cannabis.

7. Tighten the lid and store at room temperature. It can last for a month. For it to last longer, place it in the refrigerator. It can last for 3 – 5 months.

Canna-Oil (Cannabis Infused Cooking Oil)

Preparation time: 30 minutes

Cooking time: 3 hours

Makes: About 1 ½ quarts

Ingredients:

- 32 ounces cooking oil like coconut oil /olive oil/avocado oil/canola oil

- ½ ounce cannabis, decarbed

Directions:

1. Add oil into a saucepan and place it over low flame.

2. Stir in the cannabis when the oil is warm.

3. Cook for about 3 hours, stirring every 30 – 40 minutes. Make sure not to boil or simmer the oil. Turn off the heat for a few minutes if it starts boiling.

4. Place cheesecloth over a fine wire mesh strainer. Place the strainer over a large heatproof bowl.

5. Pour the oil into the strainer. Squeeze the cheesecloth to remove as much oil as possible. You can wear gloves to protect your hand from the hot oil.

6. Let the oil come to room temperature. Pour into an airtight container and use as required.

7. When cannabis is infused with olive oil, it is called canna-olive oil; when infused with coconut oil, it is called canna-coconut oil. Similarly, with the others. If a recipe calls for canna-oil, use any cannabis-infused oil that you prefer.

CBD Oil

Preparation time: 10 minutes

Cooking time: 3 hours

Makes: About 1 cup

Ingredients:

- 1 whole hemp plant (with high CBD and low THC level) chopped, ground

- 2 cups carrier oil of your choice like fractionated coconut oil, olive oil, etc.

Directions:

1. Place ground hemp plant in a canning jar. Pour carrier oil over it. Fasten the lid.

2. Pour enough water in a saucepan, about 3-4 inches in height from the bottom of the saucepan. Place a washcloth in the saucepan along with the jar.

3. Place the saucepan over medium flame.

4. When the water begins to boil, lower the heat and simmer for 3 hours. Add more water if it goes dry.

5. Shake the jar every 30 minutes. You can hold the jar with a pair of tongs to shake.

6. At the end of 3 hours, remove the saucepan from heat. Place a towel on top of the saucepan. It should be totally covered. Let it cool for 3 hours.

7. Repeat steps 3-6 once again, but this time let it cool overnight.

8. If you want stronger CBD oil, repeat steps 2-6 every day for the next 2 to 3 days.

9. When you have the oil of the preferred strength, pass the oil through cheesecloth into a dark glass bottle.

10. Store in a cool and dry area.

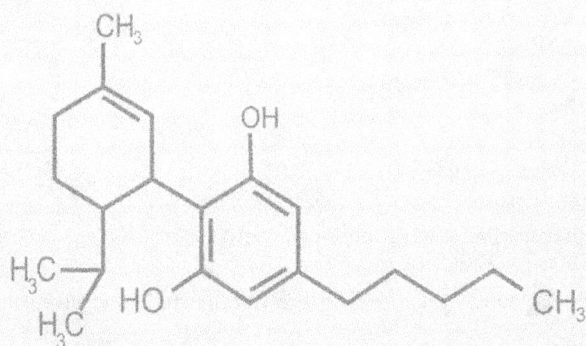

Cannabiol (CBD)

Canna-Butter

Preparation time: 5 minutes

Cooking time: 45 – 50 minutes

Makes: About ½ cup

Ingredients:

- ½ ounce cannabis buds, ground with a hand grinder into smaller pieces (do not powder it finely)

- 1 cup salted butter

Directions:

1. Place butter in a saucepan over low flame. When the butter begins to melt, add cannabis powder and stir often.

2. Let it simmer for about 50 minutes. Turn off the heat.

3. Strain into a glass container with a fitting lid. Press the residue with the back of a spoon. Discard the residue.

4. When the butter hardens, fasten the lid of the container.

5. Place the container in the refrigerator until use. If you want to make larger quantity of butter, cook for longer, approximately 60-80 minutes.

6. You can make canna- margarine or vegan butter similarly.

Canna-Honey

Preparation time: 10 minutes

Cooking time: 7 – 8 hours

Makes: About 1 ½ - 2 cups

Ingredients:

- 2 ¼ cups honey

- ½ ounce cannabis, decarbed

Directions:

1. Take a large piece of cheesecloth and place the cannabis in it. Bring the edges of the cloth together. Seal the marijuana well. It shouldn't come out of the cloth. Fasten with a string.

2. Pour honey in a crock-pot. Drop the cheesecloth bag in it.

3. Cover the pot. Cook on "Low" for 7 – 8 hours. Stir every hour.

4. Let the honey remain in the crock-pot overnight.

5. Heat the honey slightly, until just warm. Remove the cheesecloth bag and squeeze it with your hands and let the squeezed honey drop into the crockpot. Give it a good stir.

6. Store in airtight containers in a cool and dark place. It can last for 6 months.

Weed Sugar

Preparation time: 5 minutes

Cooking time: 2 hours and 30 minutes

Makes: 4 cups

Ingredients:

- 2 tablespoons citric acid (optional)

- 4 cups granulated sugar

- 2 cups cannabis tincture with grain alcohol

Directions:

1. Take a large, rimmed baking sheet and add sugar into it. Pour cannabis tincture over the sugar, stirring all the while. Add citric acid, stirring all the while.

2. Once well combined, spread the sugar evenly on the baking sheet.

3. Place the baking sheet in a preheated oven at 200°F. Point to be noted is that the oven door will remain open throughout the baking time. Also, all the windows and doors in the kitchen need to be open while baking.

4. Stir the mixture every 10-12 minutes for about 2 – 3 hours or until dry, and the sugar crystalizes once again.

5. Remove from the oven and cool completely.

6. Store in an airtight container.

Canna-Flour

Preparation time: 10 minutes

Cooking time: 0 minutes

Makes: 2 cups

Ingredients:

- 2 cups all-purpose flour
- ½ ounce cannabis, decarbed

Directions:

1. Finely grind the cannabis. Add flour and cannabis into a mixing bowl. Whisk until well incorporated. You can use a spatula or electric hand mixer or with a whisk attachment in your food processor.

2. Transfer the flour into an airtight container. Place it in a cool and dry place, like your cupboard, until use. It can last for 3 months.

Canna Milk (Cannabis Infused Milk)

Preparation time: 5 minutes

Cooking time: 45 minutes

Makes: https://theweedscene.com/cannabis-milk/ 4 – 5 cups

Ingredients:

- 1.8 ounces finely powdered cannabis

- 8 cups milk

Directions:

1. Place a pot with milk over low heat.

2. Add milk and simmer on low heat for about 35-40 minutes. Stir frequently.

3. Simmer until the milk reduces to nearly half its original quantity. The color of the milk is greenish with a tinge of yellow.

4. Strain through a fine wire mesh strainer placed over a bowl.

5. Serve hot or chilled.

Canna-Cream Cheese

Preparation time: 5 minutes

Cooking time: 30 – 40 minutes

Makes: About ¾ cuphttps://eatyourcannabis.com/canna-cream-cheese/

Ingredients:

- ½ ounce cannabis, finely ground

- ½ quart cultured buttermilk

- ½ gallon whole milk

- ¼ teaspoon salt

Directions:

1. Add milk, cannabis, and buttermilk into a saucepan. Place saucepan over medium flame. Stir occasionally until the temperature of the milk is around 170° F to 175° F on a candy thermometer.

2. Let the mixture now simmer between these temperatures for around 10 – 12 minutes. Turn off the heat when the milk curdles.

3. Place a couple of layers of cheesecloth inside a strainer. Place the strainer on a bowl.

4. Pass the curdled milk through the strainer. Retain the curds and use the milk that is remaining in some other recipe like a smoothie or discard it.

5. Cool the curds completely in the strainer.

6. Add cooled curds into a blender. Add salt and blend until smooth.

7. Transfer into an airtight container and chill until use. It can last for 3-4 days.

Cannabis Mayonnaise

Preparation time: 5 minutes

Cooking time: 0 minutes

Makes: 2 – 3 cups

Ingredients:

- 6 egg yolks
- 1 teaspoon Dijon mustard
- 2 teaspoons lemon juice
- 2 cups canna-oil cannabis-infused oil
- Salt to taste
- 2 teaspoons white vinegar
- Pepper to taste

Directions:

1. Add egg yolks, vinegar, lemon juice, salt, pepper, and Dijon mustard into a blender. Blend until smooth.

2. With the blender running blender on low speed, pour the canna-oil through the feeder tube in a thin stream until the mayonnaise is emulsified and thick.

3. If you find the mayonnaise very thick, add a teaspoon of water to dilute.

4. Transfer into an airtight container and refrigerate until use.

Cannabis Peanut Butter

Preparation time: 5 minutes

Cooking time: 0 minutes

Makes: 5 tablespoons

Ingredients:

- 3 teaspoons cannabis-infused extra-virgin olive oil

- 4 tablespoons peanut butter

Directions:

1. Add canna-oil and peanut butter (you can use either creamy or chunky peanut butter, your favorite brand) into a jar or bowl.

2. Mix well with a spoon until smooth and creamy.

3. You can spread it on bread slices or add it in smoothies or anything you like.

Cannabis-Infused Apple Cider Vinegar

Preparation time: 15 minutes

Cooking time: 0 minutes

Makes: 7 – 8 cups

Ingredients:

- 8 cups organic, unpasteurized apple cider vinegar

- 1 ½ ounces high-quality cannabis flowers

Directions:

1. Decarb the cannabis flowers, following the procedure mentioned in the first recipe.

2. Place cannabis in a large, glass gar. Pour vinegar into the jar. Fasten the lid and shake the jar constantly for about a minute.

3. Place the jar in a cool and dry area; say in a cupboard, for about 25 days. Shake the jar every 3 days.

4. Remove the lid of the jar.

5. Place cheesecloth over the rim of the jar. Fasten with a rubber band.

6. Now pour the vinegar into a pitcher. Discard the cannabis along with the cheesecloth.

7. Pour the vinegar back into the same jar.

8. Repeat steps 5 – 7 using new cheesecloth.

9. Cover the jar with the lid. Fasten the lid and refrigerate until use.

10. It can last for about 2 years in the refrigerator.

Marijuana Vinaigrette

Preparation time: minutes

Cooking time: 0 minutes

Makes: About 2/3 cup

Ingredients:

- ½ teaspoon minced garlic

- 1 teaspoon fresh basil and oregano mixture or ½ teaspoon dried basil and oregano

- ½ cup canna- oil

- Pepper to taste

- ½ tablespoon minced red onion or shallots

- 2 tablespoons balsamic vinegar or any other vinegar

- Salt to taste

Directions:

1. Place garlic, oregano or basil, onion, vinegar, salt, and pepper in a blender and blend until pureed.

2. With the blender running, pour cannabis-infused oil in a very thin stream. Blend until the vinaigrette is slightly thick and emulsified or until the consistency you desire is achieved. If your vinaigrette is not thickening, add more oil.

3. Add salt and pepper to taste.

4. Transfer into a bowl or jar. Cover and refrigerate until use. It can last for 4 – 5 days.

Italian Cannabis Dressing

Preparation time: 5 minutes

Cooking time: 0 minutes

Makes: ½ - 2/3 cup

Ingredients:

- ¼ cup canna-oil

- 2 tablespoons grated Romano cheese

- 3 tablespoons red wine vinegar

- ½ - 1 teaspoon sugar

- ½ teaspoon freshly ground pepper

- 1/8 teaspoon garlic powder

- ½ teaspoon dried basil

- ½ teaspoon dried oregano

- 1/8 teaspoon red pepper flakes

- Salt to taste

Directions:

1. Add canna-oil, cheese, vinegar, sugar, pepper, garlic powder, basil, oregano, red pepper flakes, and salt into a small glass jar.

2. Fasten the lid and shake the jar constantly for about 30 seconds.

3. Refrigerate until use. It can last for about 15 days.

4. Make sure to shake the dressing before drizzling on the salad.

Simple Salad Dressing

Preparation time: 5 minutes

Cooking time: 0 minutes

Makes: 1-½ cups

Ingredients:

- ½ cup canna-olive oil or canna- avocado oil

- 4 – 6 cloves garlic, peeled, minced

- Salt to taste

- 1 cup lemon juice

- Pepper to taste

Directions:

1. Add oil, garlic, salt, lemon juice, and pepper into a mason's jar.

2. Fasten the lid and shake the jar constantly for about 30 seconds.

3. Refrigerate until use. It can last for about 15 days.

4. Make sure to shake the dressing before drizzling on the salad. You can also top it on meat.

Sweet Salad Dressing

Preparation time: 5 minutes

Cooking time: 0 minutes

Makes: About ¾ cup

Ingredients:

- 1 tablespoon lemon juice

- Salt to taste

- 2/3 cup mashed berries of your choice

- 3 tablespoons canna-olive oil or canna-avocado oil

- Pepper to taste

Directions:

1. Add oil, berries, salt, lemon juice, and pepper into a mason's jar.

2. Fasten the lid and shake the jar constantly for about 30 seconds.

3. Refrigerate until use. It can last for about 2 days.

4. Make sure to shake the dressing before drizzling on the salad. You can also top it on meat.

Honey Mustard Salad Dressing

Preparation time: 5 minutes

Cooking time: 0 minutes

Makes: 1 – 1 1/3 cups

Ingredients:

- ½ cup apple cider vinegar

- 2/3 cup canna-olive oil or canna- avocado oil

- 3 – 4 teaspoons Dijon mustard

- Honey to suit your taste (you can also use canna-honey if you want more cannabis in your diet)

Directions:

1. Add vinegar, canna-oil, mustard, and honey into a bowl and whisk until well combined and smooth.

2. Cover the bowl. Refrigerate until use. It can last for about 5 – 6 days.

3. Make sure to stir the dressing before drizzling on the salad. You can also top it on meat.

Cannabis-Infused BBQ Sauce

Preparation time: 10 minutes

Cooking time: 15 minutes

Makes: 7 – 8 cups

Ingredients:

- 4 cups ketchup

- 1 cup apple cider vinegar

- 2/3 cup sugar

- 2/3 cup light brown sugar

- 1 tablespoon onion powder

- 2 tablespoons Worcestershire sauce

- ½ cup canna-oil

- 2 cups water

- 1 tablespoon freshly ground pepper

- 1 tablespoon ground mustard

Directions:

1. Add ketchup, vinegar, sugar, light brown sugar, onion powder, Worcestershire sauce, canna-oil, water, pepper, and mustard into a pot.

2. Place the pot over low flame. Stir often until sugars dissolve completely. Let it cook for about 18 – 20 minutes.

3. Turn off the heat. Transfer into an airtight container.

4. You can use it for BBQ chicken or meat as a marinade or use it as a dip or spread it over cooked meat as topping.

Hazelnut Spread

Preparation time: 10 minutes

Cooking time: 15 minutes

Makes: About a cup

Ingredients:

- 5.3 ounces cream

- 0.6 – 1 ounce canna-butter (it depends on the strength of the cannabis)

- 1 teaspoon organic hazelnut extract, at room temperature

- 5 ounces semi-sweet chocolate chips

- 2.8 ounces hazelnut butter, at room temperature

Directions:

1. Add cream and hazelnut butter into a saucepan. Place the saucepan over medium flame. Whisk using a hand whisk until well combined.

2. When the mixture begins to simmer, turn off the heat.

3. Place chocolate chips in a bowl. Pour hazelnut –cream mixture over the chocolate chips. Cover and let it rest for 3 – 4 minutes.

4. Once again, whisk the mixture using a hand whisk until smooth and chocolate chips dissolve completely.

5. Add hazelnut extract and canna-butter and whisk until smooth and well combined.

6. Continue whisking until butter melts.

7. Cover the spread with cling wrap, with the cling wrap touching the spread.

8. Chill until use.

9. To use: Remove the bowl from the refrigerator and keep it on your countertop for an hour before serving.

Cannabis Salsa

Preparation time: 15 minutes

Cooking time: 0 minutes

Makes: about 1 ½ cups

Ingredients:

- 1 cup chopped fresh tomatoes

- ¼ cup diced onion

- A handful fresh cilantro, chopped

- ¼ fresh jalapeño, diced

- Lime juice to taste

- Salt to taste

- ¼ cup diced bell pepper

- 2 tablespoons canna-oil

- Pepper to taste

Directions:

1. Add tomatoes, onion, cilantro, jalapeño, lime juice, bell pepper, canna-oil, and pepper into a bowl and toss well.

2. Cover and chill until use. It can last for 2 days.

Cannabis Caramel Sauce

Preparation time: 5 minutes

Cooking time: 30 – 40 minutes

Makes: 8 – 10 servings

Ingredients:

- 2 cans (14 ounces each) full-fat coconut milk
- 1 cup coconut sugar
- 4 teaspoons canna-coconut oil
- 1 teaspoon vanilla extract
- Salt to taste (optional)

Directions:

1. Add coconut sugar and coconut milk into a saucepan. Place the saucepan over medium flame.

2. Stir frequently until the mixture comes to a simmer. Stir frequently until thickened to the consistency you desire. Point to be noted is that the sauce will become thicker on cooling.

3. Turn off the heat and cool. If the consistency you desire is achieved on cooling, go to the next step else simmer for some more time.

4. Whisk in salt, canna-coconut oil, and vanilla. Keep whisking until oil is well combined.

5. Pour into a container and refrigerate until use.

Cannabis Glycerin Concentrate

Preparation time: 5 minutes

Cooking time: 24 – 26 hours

Makes: About ¾ cup

Ingredients:

- 1.8 ounces cannabis trim

- 1 cup food grade vegetable glycerin

Directions:

1. Decarb the cannabis as given in the first recipe. Roughly powder it and add into a bowl along with glycerin and stir.

2. Place the bowl in the slow cooker. Set the temperature to 180° F and time for 24 – 26 hours, stirring after every 3 hours. Use a wooden spoon to stir.

3. Take a large piece of cheesecloth and place it over a jar or storage container. Pour the glycerin concentrate into it. Bring together the edges and squeeze out as much glycerin as possible. Fasten the lid and chill until use.

4. You can use it instead of honey to sweeten smoothies etc.

CHAPTER 4

BREAKFAST RECIPES

Weed Crepes

Preparation time: 10 minutes

Cooking time: 10 minutes

Makes: 4 servings

Ingredients:

- 2/3 cup canna milk

- ½ cup all-purpose flour

- 1 tablespoon white sugar

- 2 eggs, lightly beaten

- 1 tablespoon butter, melted + extra to fry

- ¼ teaspoon salt

Directions:

1. Combine canna-milk, flour, sugar, eggs, butter, and salt in a bowl and whisk until you get a batter that is smooth and free from lumps.

2. Place a medium-sized pan over medium flame. Add a little butter or oil.

3. Pour 3 tablespoons of the batter into the pan. Swirl the pan to spread the batter thinly. Cook until the underside is golden brown. Flip the crepe over and cook the other side until golden brown.

4. Remove the crepe onto a plate.

5. Repeat steps 2 – 4 and make the remaining crepes.

Cannabis Waffles

Preparation time: 10 minutes

Cooking time: 20 minutes

Makes: 2 – 3 servings

Ingredients:

- 2 tablespoons canna-butter

- 6 tablespoons sugar

- 1 large egg, separated

- ½ cup butter, melted

- 1 cup all-purpose flour

- 1 ¾ teaspoons baking powder

- ¾ cup whole milk

- ½ teaspoon vanilla extract

To serve:

- Fresh berries

- Syrup of your choice

- Whipped cream

- Any other toppings of your choice

Directions:

1. Add sugar, flour, and baking powder into a bowl and stir until well combined.

2. Beat the yolk in a bowl. Beat in the butter, canna-butter, milk, and vanilla.

3. Pour into the bowl of dry ingredients and stir until just incorporated, making sure not to overbeat.

4. Beat whites with an electric hand mixer until stiff peaks are formed.

5. Add the whites into the batter and fold gently.

6. Set up your waffle iron and preheat it following the manufacturer's instructions. Grease the iron with some butter or oil.

7. Pour some batter into the waffle iron (about ½ cup). Set the timer for 4 – 6 minutes or until the way you like it cooked. Remove the waffle and serve with any of the suggested serving options.

8. Repeat steps 6 – 7 and make the remaining waffles.

Weed French Toast

Preparation time: 10 minutes

Cooking time: 45 minutes

Makes: 8 servings, 2 toasts each

Ingredients:

- 2 French baguettes, cut each into 8 slices, crosswise, ¾ - 1 inch thick

- 6 tablespoons canna-butter

- 3 tablespoons butter, unsalted + extra for greasing

- 8 eggs

- 1/3 cup sugar

- 1 ½ cups milk

- 6 tablespoons maple syrup

- 2 teaspoons salt

- 2 teaspoons vanilla extract

- Powdered sugar to top

Directions:

1. Prepare a baking dish by greasing it with butter.

2. Add canna-butter and butter into a bowl and mix well. Spread this mixture on one side of each slice of bread.

3. Place the bread slices in the baking dish with the buttered sided facing up.

4. Whisk together eggs, sugar, milk, maple syrup, salt, and vanilla extract in a bowl until well combined.

5. Pour this mixture over the bread slices. Press it down with a spoon.

6. Cover and chill overnight.

7. Place the baking dish in an oven that has been preheated to 350° F and bake for about 45 minutes or until golden brown.

8. Remove the baking dish from oven. Cool for 5 minutes.

9. Sprinkle powdered sugar on top and serve.

Weed Pancakes

Preparation time: 10 minutes

Cooking time: 10 minutes

Makes: 8 servings

Ingredients:

- 2 tablespoons white sugar
- ½ teaspoon baking soda
- 1 ½ cup canna-milk
- 2 cups all-purpose flour
- 2 teaspoons baking powder
- 1 teaspoon salt
- 1 tablespoon canna butter, melted
- 1 tablespoon white vinegar
- Cooking spray

To serve:

- Honey or any other syrup to serve
- Berries to serve
- Whipped cream
- Any other toppings of your choice

Directions:

1. Mix together all the dry ingredients, i.e., flour, baking soda, baking powder, salt, and sugar, in a bowl.

2. Pour milk and vinegar into another bowl and let it sit for a few minutes, about 5 – 8 minutes.

3. Add eggs and butter and whisk well.

4. Pour butter - milk mixture into the bowl of dry ingredients and whisk well until it is smooth and free from lumps. Set aside for 10-15 minutes.

5. Place a nonstick skillet over medium heat. Spray with cooking spray.

6. Pour about 4 tablespoons of batter, you can use a ladle.

7. In a couple of minutes, bubbles will be visible on the top of the pancake. Cook until the underside is golden brown. Turn the pancake over and cook the other side as well.

8. Remove pancakes from the pan and keep warm.

9. Repeat steps 5 – 8 and make the remaining pancakes.

10. Serve with any of the suggested toppings.

Cannabis Granola Breakfast Bars

Preparation time: 10 minutes

Cooking time: 20 – 30 minutes

Makes: 20 – 25 servings

Ingredients:

- 1 cup canna-coconut oil (infused coconut oil), melted

- 2 cups chopped nuts

- 2 teaspoons baking soda

- 3 teaspoons ground cinnamon

- Flavoring of your choice (optional)

- 1 cup brown flaxseed meal

- 1 cup honey or maple syrup

- 1/8 teaspoon salt

- 6 cups oatmeal

- 1 cup berries or fruits of your choice, fresh or dried

Directions:

1. Prepare a large rimmed baking sheet by lining it with parchment paper. Also, prepare the oven by preheating it to 300° F.

2. Add oatmeal, flaxseed meal, nuts, and cinnamon into a bowl and mix well.

3. Add oil, salt, honey, and any flavoring, if using, into a bowl and whisk until well incorporated. Pour into the bowl of the oatmeal mixture.

4. Mix until well incorporated. Transfer the mixture onto the prepared baking sheet. Spread it evenly but do not press the mixture onto the baking sheet.

5. Place the baking sheet in the oven and bake for 20-30 minutes or until golden brown. Stir once halfway through baking.

6. Remove the baking sheet from the oven and mix in the berries. Spread the mixture again on the baking sheet, evenly.

7. Let it cool slightly. Make 20 – 25 equal portions of the mixture and shape into bars.

8. Transfer into an airtight container and refrigerate until use. It can last for 5 – 6 days.

Blueberry Muffins with Weed Streusel

Preparation time: 20 minutes

Cooking time: 30 minutes

Makes: 18 servings

Ingredients:

For muffins:

- 2 cups milk

- 8 tablespoons butter, at room temperature

- 2 eggs

- 4 2/3 cups flour

- A large pinch salt

- 2/3 cup sugar

- 5 teaspoons baking powder

- 2 teaspoons vanilla extract

- 3 cups frozen or fresh blueberries

For streusel:

- 2/3 cup all-purpose flour

- ½ cup chilled canna-butter

- 1 cup sugar

- 1 teaspoon ground cinnamon

Directions:

1. To prepare muffin pans and oven: Take 3 muffin pans of 6 counts each. Grease them with cooking spray. Place cupcake papers in the muffin pans.

2. Place rack in the center of the oven and preheat the oven to 350° F.

3. To make muffins: To mix the dry ingredients: Mix together flour, salt, and baking powder into a mixing bowl and stir well.

4. To mix wet ingredients: Add sugar and butter into a large mixing bowl and whisk with an electric hand mixer until creamy.

5. Add vanilla and mix well. Add eggs, one at a time, and whisk well each time.

6. Add dry ingredients into the bowl of wet ingredients. Also, pour the milk and mix until just combined, making sure not to overbeat.

7. Add blueberries and stir well.

8. Pour into the muffin molds, up to 2/3 of the molds.

9. To make streusel topping: Add flour, sugar, and cinnamon into a bowl and mix well.

10. Add canna butter and mix it well into the mixture of flour until crumbly in texture. Scatter this mixture over the batter in the muffin cups.

11. Place the muffin pans in the oven and bake the muffins for about 20 – 25 minutes or until light brown. Cook in batches if required.

12. Cool completely on a cooling rack and serve.

13. Blueberries can be replaced with strawberries or any other berries of your choice.

Weed Breakfast Casserole

Preparation time: 15 minutes

Cooking time: 45 minutes

Makes: 6 servings

Ingredients:

- ½ package (from a 16 ounces package) ground pork breakfast sausage

- ½ can (from a 10.75 ounces can) condensed cream of mushroom soup

- ½ can (from a 4.5 ounces can) sliced mushrooms

- ¼ cup shredded cheddar cheese

- 6 eggs

- ½ + 1/8 cup canna-milk

- Salt to taste

- ½ package (from a 32 ounces package) frozen potato rounds

- Pepper to taste

Directions:

1. Place a skillet over medium-high flame. Add sausage and cook until brown, stirring often.

2. Meanwhile, prepare a baking dish by greasing it with some cooking spray. Prepare the oven by preheating it to 350° F. Place rack in the center of the oven.

3. Add eggs, milk, and mushroom soup into a bowl and whisk until well incorporated.

4. Add mushrooms and sausage and stir well.

5. Place the potato rounds in the baking dish. Spoon the sausage mixture into the baking dish and place the dish in the oven. Bake until set and light brown on

top. It should take around 35 – 40 minute. Anyway, keep a watch over it after 30 minutes of baking.

6. Scatter cheese on top and place the baking dish back in the oven. Bake for another 5 to 7 minutes or until cheese melts.

7. Serve hot.

Breakfast "Baked" Burritos

Preparation time: 10 minutes

Cooking time: 15 minutes

Makes: 2 servings

Ingredients:

- 3 ounces bacon

- 2 eggs

- 1.5 ounces cheddar cheese, shredded

- 1 tablespoon canna-butter

- 6-7 tablespoons refried beans

- 2 flour tortillas (10 inches each)

Directions:

1. Place a large deep skillet over medium-high heat. Add bacon and cook until brown.

2. Remove with a slotted spoon and place on a plate lined with paper towels. Chop the bacon into smaller pieces.

3. Warm the tortillas following the instructions on the package.

4. Place a nonstick skillet over medium heat. Add canna-butter.

5. When the butter melts, crack the eggs and fry the eggs according to your preference.

6. Spread refried beans over the warm tortillas. Place half the pieces of bacon and an egg on each tortilla.

7. Sprinkle cheese on top.

8. Wrap like a burrito and serve.

Veggie Frittata

Preparation time: 10 minutes

Cooking time: 20 minutes

Makes: 4 servings

Ingredients:

- 4 large eggs

- Sea salt to taste

- 1 tablespoon cannabis-infused olive oil or hash oil

- 1 small onions, thinly sliced

- 2 ounces feta cheese, crumbled

- 3 tablespoons milk

- Pepper to taste

- ½ medium red bell pepper, thinly sliced

- 1 cup baby spinach leaves

- Fresh basil leaves for garnishing (optional)

Directions:

1. Whisk the eggs well. Add milk, salt, and pepper. Whisk until well combined.

2. Add cannabis-infused olive oil to an ovenproof skillet. Place the skillet over medium flame.

3. When oil is heated, add onions and bell pepper and sauté until slightly tender.

4. Stir in the spinach and sauté until the spinach wilts.

5. Pour the egg mixture over the onion mixture. Cook for about a minute.

6. Sprinkle feta cheese on top of the egg layer, making sure not to stir.

7. When the sides are cooked and the middle undercooked, turn off the heat.

8. Shift the skillet into an oven that has been preheated to 350° F and bake for about 15 minutes. Set the oven to broil mode and bake for another 5 minutes or until or until very light golden brown on top.

9. Take out the skillet from oven. Cool for 5 minutes.

10. Garnish with fresh basil leaves. Cut into wedges and serve.

Kale and Chevre Goat Cheese Frittata

Preparation time: 15 minutes

Cooking time: 25 minutes

Makes: 4 servings

Ingredients:

- 4 large eggs

- Sea salt to taste

- 1 ½ tablespoons cannabis-infused olive oil or canna-butter

- ½ large shallot, thinly sliced

- 3 mushrooms, sliced

- 2 small cloves garlic, peeled, thinly sliced

- 3 tablespoons crumbled soft goat's cheese

- Pepper to taste

- 1 cup baby kale leaves

- 4 tablespoons almond milk

- ½ small red onion, chopped

Directions:

1. Add ½ tablespoon cannabis-infused olive oil to an ovenproof skillet. Place the skillet over medium flame.

2. When oil is heated, add shallot, onion, and garlic and sauté for a minute.

3. Stir in the mushrooms and cook until slightly tender.

4. Stir in the kale and sauté until the kale wilts. Remove the vegetables onto a plate. Clean the skillet with a paper napkin and place it back over medium flame.

5. Add remaining oil into the skillet and let it heat. Swirl the pan to spread the oil.

6. Whisk the eggs well. Add almond milk, salt, and pepper. Whisk until well combined.

7. Pour the egg mixture into the skillet. Scatter the onion mixture all over the egg. Cook for about a minute.

8. Sprinkle goat's cheese on top of the vegetables, making sure not to stir.

9. When the sides are cooked and the middle undercooked, turn off the heat.

10. Shift the skillet into an oven that has been preheated 350° F and bake for about 15 minutes. Set the oven to broil mode and bake for another 5 minutes or until or until very light golden brown on top.

11. Take out the skillet from oven. Cool for 5 minutes.

12. Cut into wedges and serve.

Weed Omelet

Preparation time: 5 minutes

Cooking time: 10 minutes

Makes: 2 servings

Ingredients:

- 8 large eggs

- ½ teaspoon pepper or to taste

- ¼ cup chopped red bell pepper

- ¼ cup chopped green onion

- ½ cup shredded cheese

- ½ teaspoon salt or to taste

- ¼ cup cooked diced ham or any other meat of your choice

- 2 tablespoons butter

- 1 cup canna-milk

Directions:

1. Add eggs into a bowl and whisk well. Add milk, salt, and pepper and whisk well.

2. Stir in the ham, bell pepper, and green onion.

3. Place a skillet over medium flame. Add 1 tablespoon of butter and let it melt. When butter melts, pour half the egg mixture.

4. Tilt the pan to spread the egg mixture.

5. Cook until the underside is golden brown. Turn the omelet over and cook the other side until golden brown.

6. Carefully remove the omelet onto a plate and serve.

7.   Repeat steps 3 – 6 and make the other omelet.

Banana Nut Bread

Preparation time: 15 minutes

Cooking time: 60 – 90 minutes

Makes: 2 loaves

Ingredients:

For dry ingredients:

- 2 teaspoons ground cinnamon

- 2 teaspoons baking powder

- 2 teaspoons baking soda

- 3 cups flour

- 1 cup whole wheat flour

- 2 teaspoons baking powder

Other ingredients:

- 1 cup brown sugar

- 1 cup granulated sugar

- 2 eggs

- 2 teaspoons milk

- 1 cup chocolate chips

- 1 cup canna-butter

- 6 bananas, mashed

- 1 teaspoon vanilla extract

- 1 cup chopped walnuts

Directions:

1. Add canna-butter, brown sugar, and sugar to a large mixing bowl and beat with an electric hand mixer until creamy.

2. Add eggs at a time and beat well each time. Scrape the sides of the bowl, time to time.

3. Beat until the mixture is light and creamy. Set aside for a while.

4. To mix dry ingredients: Sift together flour, whole-wheat flour, baking powder, baking soda, and cinnamon.

5. Add bananas, milk, and vanilla into another bowl and beat until well combined. Pour into the mixing bowl.

6. Whisk until well combined.

7. Add the mixture of dry ingredients and mix until well combined and free from lumps.

8. Add the walnuts and chocolate chips and fold gently.

9. Grease 2 loaf pans with some cooking spray. Divide the batter among the loaf pans.

10. Place the baking pan in an oven that has been preheated to 325° F and bake for 1 - 1 ½ hours or until brown on top.

11. Cool on your countertop for 15 minutes.

12. Invert on to a cooling rack. Cool for some more time.

13. Slice and serve. Store in an airtight container in the refrigerator.

Berry and Banana Smoothie

Preparation time: 5 minutes

Cooking time: 0 minutes

Makes: 1 – 2 servings

Ingredients:

- ½ cup coconut milk or almond milk, unsweetened

- ½ medium bananas, peeled, sliced

- ¾ cup frozen berries of your choice

- ½ – 1 tablespoon canna-coconut oil or any other canna-cooking oil

- 2 teaspoons chia seeds or ½ tablespoon hemp seeds (optional)

- ¼ cup orange juice

Directions:

1. Place bananas, berries, and chia seeds in a blender.

2. Pour milk, canna – oil, and orange juice. Blitz for 30 – 40 seconds or until smooth and creamy.

3. Pour into 1 – 2 glasses. Serve immediately with crushed ice.

Green Juice

Preparation time: 10 minutes

Cooking time: 0 minutes

Makes: 2 servings

Ingredients:

- 10 handfuls spinach

- 14 large cannabis fan leaves

- 1 lemon, peeled

- 6 kale leaves, torn

- 1 cucumber, chopped into chunks

- 2 Fuji apples, chopped into chunks

Directions:

1. Juice together spinach, cannabis fan leaves, lemon, kale, cucumber, and apples in a juicer.

2. Pour into 2 glasses and serve.

Sweet n Spicy Juice

Preparation time: 10 minutes

Cooking time: 0 minutes

Makes: 2 servings

Ingredients:

- 6 handfuls spinach

- 40 small sugar leaves

- 20 large cannabis fan leaves

- 1 jalapeño, chopped

- 10 kale leaves, torn

- 2 cucumbers, chopped into chunks

- 2 cups pineapple chunks

- 2 large cannabis buds

Directions:

1. Juice together spinach, sugar leaves, cannabis fan leaves, jalapeño, kale, cucumber, cannabis buds, and pineapple in a juicer.

2. Pour into 2 glasses. Add water to dilute if desired and serve with crushed ice.

Strawberry Banana Smoothie

Preparation time: 5 minutes

Cooking time: 0 minutes

Makes: 2 servings

Ingredients:

- 6 tablespoons canna-butter

- 1 cup frozen strawberries

- 1 cup thin vanilla yogurt

- Ice cubes, as required

- 2 bananas, sliced, frozen

- 6 tablespoons shredded coconut

- ½ cup coconut milk or milk of your choice

Directions:

1. Add canna-butter, strawberries, yogurt, ice cubes, bananas, coconut, and milk into a blender.

2. Blitz for 30 – 40 seconds or until smooth and creamy.

3. Pour into 2 glasses and serve

Raw Cannabis Green Smoothie

Preparation time: 10 minutes

Cooking time: 0 minutes

Makes: 2 – 3 servings

Ingredients:

- 10 handfuls spinach

- 6 kale leaves

- 14 large cannabis fan leaves

- 1 cucumber, chopped

- 2 Fuji apples, peel if desired, cored, chopped

- Juice of a lemon

Directions:

1. Add spinach, kale, cannabis, cucumber, apples, and lemon juice into a blender and blitz until smooth.

2. Pour into 2 – 3 glasses and serve.

Weed Iced Coffee

Preparation time: 2 minutes

Cooking time: 1 minute

Makes: 4 servings

Ingredients:

- 1 cup water

- 2 cups ice cubes or more if desired

- 1 cup canna- milk

- 4 teaspoons instant coffee granules

- 1 can (5 ounces) sweetened condensed milk

- 2 tablespoons chocolate syrup

Directions:

1. Pour water into a saucepan. Heat the water until warm. Turn off the heat.

2. Add instant coffee and stir until it dissolves.

3. Pour into a blender. Add ice cubes, canna milk, chocolate syrup, and condensed milk and blitz until well combined.

4. Pour into 4 glasses and serve.

Indian Bhang

Preparation time: 5 minutes

Cooking time: 2 – 3 minutes

Makes: 4 servings

Ingredients:

• ¼ ounce butter

• 0.07 ounce cannabis or hash

• 4 cups milk

• ¼ teaspoon ground cinnamon or ground nutmeg

• Honey or sugar to taste (optional)

Directions:

1. Add butter into a saucepan. Place saucepan over low heat. Add cannabis and let it cook for a minute in the melted butter.

2. Add milk and the chosen spice and stir. When milk is warm, turn off the heat.

3. Add honey or sugar to taste if desired.

4. You can serve as it is or with vodka.

CHAPTER 5

DINNER RECIPES

Main Course

Cannabis-Infused Butter Chicken

Preparation time: 4 hours

Cooking time: 30 minutes

Makes: 3 servings

Ingredients:

To marinate chicken:

- 1 pound boneless, skinless chicken breasts, cubed

- ¼ teaspoon freshly cracked pepper

- ½ teaspoon turmeric powder

- ¼ teaspoon sea salt or to taste

- ½ teaspoon chili powder

- ¼ cup Greek yogurt

For butter gravy:

- 2 tablespoons canna-butter

- 1 tablespoon butter

- 1 large onion, diced

- ½ teaspoon chili powder

- ½ teaspoon turmeric powder

- ½ tablespoon garam masala

- ½ teaspoon ground cumin

- ½ teaspoon cayenne pepper

- ¼ teaspoon pepper

- 2 cloves garlic, peeled, minced

- ½ tablespoon brown sugar

- ¼ cup water

- Salt to taste

To serve:

- A handful fresh cilantro, chopped

- Cooked, long grain rice

Directions:

1. To marinate chicken: Add yogurt, salt, and all the spices into a bowl and stir well.

2. Add chicken and stir until chicken is well coated with the marinade.

3. Cover and chill for 4 hours or longer if you have time on hand.

4. Place a skillet over medium flame. Add butter and let it melt.

5. Remove only chicken from the marinade and shake to drop off excess marinade. Place the chicken in the skillet and cook until brown all over.

6. Stir in the marinade and transfer into a bowl.

7. To make sauce: Add a tablespoon of canna-butter into the skillet.

8. When butter melts, add onion and sauté for a couple of minutes.

9. Stir in the ginger and garlic and sauté for about a minute until you get a nice aroma.

10. Stir in all the spices, salt, and brown sugar and cook for a few seconds.

11. Stir in tomato sauce and water and let it come to a boil.

12. Now add cream and stir. Let it come to a boil. Add chicken back into the pot along with the drippings and cook for 20 minutes or until chicken is cooked through.

13. Sprinkle salt and pepper to taste. Add a tablespoon of canna-butter and stir.

14. Sprinkle cilantro on top. Serve butter chicken over rice.

Cannabis Chicken Fajitas

Preparation time: 15 minutes

Cooking time: 30 minutes

Makes: 4 servings

Ingredients:

- 1 tablespoon canna-oil

- ½ teaspoon ground cumin

- ½ pound skinless, boneless chicken breasts

- 1 bell pepper, sliced

- 1 clove garlic, finely chopped

- 4 corn tortillas

- 3 tablespoons crumbled cotija cheese

- ½ teaspoon chili powder

- Freshly ground pepper to taste

- 1 tablespoon extra-virgin olive oil

- ½ red onion, sliced

- ¼ teaspoon grated lime zest

- Lime juice to taste

- ¼ cup prepared Pico de Gallo

Directions:

1. Combine ¼ teaspoon cumin, pepper, and salt to taste in a bowl. Sprinkle this mixture over the chicken and rub it well into it.

2. Place a large skillet over medium-high flame. Add ½ tablespoon extra-virgin olive oil. Once oil is hot, add chicken and cook until golden brown.

3. Remove chicken from the pan and place on a baking sheet.

4. Place the baking sheet into an oven than has been preheated to 350° F and bake for 10 - 15 minutes or until well-cooked.

5. Remove the chicken from the baking sheet and place on your cutting board. When cool enough to handle, cut into slices and keep warm.

6. Add ½ tablespoon extra-virgin olive oil into the skillet. Stir in onion, garlic, bell pepper, and cook for a couple of minutes.

7. Add remaining cumin and cook until vegetables are light brown.

8. Add canna-oil, lime zest, and a sprinkle of water. Add salt to taste. Cook for a couple of minutes and turn off the heat.

9. Heat the tortillas according to the instructions on the package.

10. Divide the chicken slices over the tortillas. Divide the vegetable mixture, cheese, and Pico de Gallo over the vegetables. Drizzle lime juice on top and serve.

Easy Cannabis-Infused Chicken Adobo

Preparation time: 15 minutes + marinating time

Cooking time: 30 minutes

Makes: 3 servings

Ingredients:

For marinade:

- 1 pound fresh chicken legs

- 1 teaspoon garlic powder

- 2 tablespoons soy sauce

- ½ teaspoon grated, fresh ginger

- 1 teaspoon freshly ground black pepper

- 1 tablespoon coconut oil

For sauce:

- 2 tablespoons canna-coconut oil

- 2 small bay leaves

- ½ medium onion, sliced

- 2 tablespoons apple cider vinegar

- ½ teaspoon maple syrup

- ¾ cup water

- 1 teaspoon whole peppercorns

- 2 cloves garlic, crushed

- 1 teaspoon brown sugar or to taste

- ½ teaspoon sea salt

Directions:

1. To marinate the chicken: Add soy sauce, garlic powder, ginger, and pepper into a bowl and stir.

2. Add chicken and stir until chicken is well coated with the marinade.

3. Cover and place it in the refrigerator for 1 – 8 hours.

4. To cook chicken: Place a pan over medium flame. Add oil. When oil melts and is well heated, place the chicken in the pan without the marinade.

5. Cook until brown all over.

6. To make sauce: Now add the marinade and water and let it come to a boil.

7. Add onion, garlic, peppercorns, and bay leaves and stir.

8. Cook on low until chicken is cooked through.

9. Stir in sugar, vinegar, salt, and maple syrup. Mix well and continue simmering for 5 – 8 minutes or until thick and you can see some oil. Stir every 3 – 4 minutes.

Chicken Pot-Cacciatore

Preparation time: 15 minutes

Cooking time: 50 – 60 minutes

Makes: 8 – 10 servings

Ingredients:

- 2 fryer chickens, with skin, cut into pieces, rinsed,

- 2 tablespoons canna-butter

- 2 tablespoons olive oil

- Pepper to taste

- 1 glass white wine (optional)

- ¼ cup whole green olives

- ¼ cup whole black olives

- Salt to taste

- 2 large onions, cut into ½ inch wedges

- 2 – 3 cups small cremini mushrooms

Directions:

1. Pat the chicken with paper towels until absolutely dry.

2. Season the chicken pieces with salt and pepper.

3. Place a large skillet over medium flame. Add oil as well as canna- butter.

4. Add chicken pieces once the butter melts and cook until brown all over.

5. Take out the chicken using a slotted spoon and place on a plate lined with paper towels. Do not discard the fat.

6. Add onions into the same pan and cook until slightly soft.

7. Add chicken and mushrooms and cook for 4-5 minutes.

8. Stir in wine and simmer for a couple of minutes. Add green olives and black olives and mix well. Turn off the heat. Cover and set aside for a few minutes before serving.

Cannabis-Infused Thai Green Curry

Preparation time: 15 minutes

Cooking time: 20 – 25 minutes

Makes: 2 – 3 servings

Ingredients:

For green curry paste:

- 3 green chilies, deseeded, chopped

- ½ inch ginger, peeled, grated

- ½ bunch coriander (roots, stalks, and leaves), rinsed well, chopped

- Juice of ½ lime

- Zest of ½ lime grated

- ½ inch galangal, peeled, chopped

- 1 shallot, roughly chopped

- 1 clove garlic, crushed

- 1 stalk fresh lemongrass, chopped

- 4 kaffir lime leaves or use extra lemon zest

- ½ tablespoon coriander seeds, crushed

- ½ teaspoon whole peppercorns, crushed

- ½ teaspoon ground cumin

- 1 teaspoon Thai fish sauce

- 1 ½ tablespoons olive oil

For curry:

- 2 medium potatoes, peeled, chopped into chunks

- ½ cup cubed zucchini

- 2 small cloves garlic, peeled, minced

- ½ cup diced red bell pepper

- 7 – 8 green beans or snap peas, trimmed, halved

- ½ tablespoon olive oil

- Thai green curry paste to suit your taste

- 1 cup coconut milk

- 1 cup chicken broth

- ½ teaspoon brown or cane sugar

- ½ pound boneless chicken, cut into bite-size pieces

- 1 kaffir lime leaf, thinly sliced or 2 strips lemon zest

- ½ Thai chili, sliced

- 2 teaspoons canna-butter

- ½ teaspoon vegetable oil

- Steamed rice to serve

Directions:

1. To make green curry paste: Add chilies, ginger, galangal, cilantro, garlic, lemongrass, shallot, lemongrass, kaffir leaves, spices, fish sauce, and oil into a blender and blend until smooth. You can also pound in a mortar and pestle.

2. To make curry: Cook potatoes in a pot of boiling water for 5 minutes or until fork-tender. Add zucchini, beans, and bell pepper.

3. Drain off after 3 minutes and set it aside.

4. Place a large pan over medium flame. Add oil and let it heat. Once oil is heated, add garlic and stir until it turns light golden brown in color.

5. Add curry paste, 2 – 4 teaspoons, or more if desired and stir-fry for about 20 seconds or until you get a nice aroma.

6. Add coconut milk and canna-butter and let it melt.

7. Add broth and stir. Add more broth if you want more gravy.

8. Add fish sauce, brown sugar, Thai chili, and chicken and stir. Lower the heat and cover with a lid. Simmer until chicken is cooked through.

9. Add cooked vegetables and heat thoroughly.

10. Add basil and turn off the heat. Garnish with kaffir leaves and serve over rice.

Classic Cannabis Lasagna

Preparation time: 20 minutes

Cooking time: 45 – 50 minutes

Makes: 2 – 3 servings

Ingredients:

- ½ pound ground turkey or beef

- ½ medium onion, finely chopped

- ½ can (from a 14.5 ounces can) stewed tomatoes

- ½ can (from a 6 ounces can) tomato paste

- 1 large egg, beaten

- ¼ cup ricotta cheese

- 1 teaspoon chopped fresh parsley

- ½ teaspoon pepper or to taste

- 4 ounces shredded cheddar cheese

- 1 ½ tablespoons canna-extra-virgin olive oil

- 1 clove garlic, minced

- ½ jar (from a 6 ounce jar) tomato sauce

- ½ box (from an 8 ounces bag) no-boil lasagna noodles

- ¾ cup cottage cheese

- 4 ounces shredded mozzarella cheese

- 4 ounces grated parmesan cheese

- 1 teaspoon salt

Directions:

1. Take rectangular baking dish and spray some oil in the dish.

2. Place a skillet over medium-low heat. Add canna-oil and heat slightly, making sure not to let the oil smoke.

3. Add garlic and onion and sauté for a minute. Add ½ teaspoon salt and ¼ teaspoon pepper. Mix well.

4. Stir in turkey and cook until brown. As the turkey is cooking, break it using a wooden spoon. Discard excess fat from the pan.

5. Add tomato sauce, stewed tomatoes, and tomato paste into the pan and stir. Cover and cook on low for about 10 minutes. Stir every 4 – 5 minutes.

6. Add egg, cottage cheese, ¼ cup parmesan cheese, ricotta cheese, parsley, ½ teaspoon salt, and ¼ teaspoon pepper into a bowl and mix until well incorporated.

7. Add a little of the turkey sauce into the baking dish and spread it in a thin layer.

8. Place a layer of lasagna noodles in the baking dish, overlapping by about ½ an inch.

9. Spread half the egg- cheese mixture over the noodles, followed by half the remaining mozzarella and half the cheddar cheese. Spread some more turkey sauce over this layer.

10. Repeat a layer of lasagna noodles. Spread remaining half egg-cheese mixture over the noodles, followed by half the remaining mozzarella and half the cheddar cheese. Spread some more turkey sauce over this layer.

11. Top with remaining parmesan cheese.

12. Place the baking dish in an oven that has been preheated to 350° F and bake for about 25-30 minutes or until you can see the sauce bubbling.

13. Fresh bread tastes great with this lasagna.

Cannabis-Infused Turkey Bolognese

Preparation time: 15 minutes

Cooking time: 30 minutes

Makes: 3 – 4 servings

Ingredients:

- 4-6 drops cannabis tincture

- Pepper to taste

- ½ cup grated parmesan cheese

- Salt to taste

- 2 tablespoons chopped parsley

- 2 cloves garlic, peeled, minced

- 1 stalk celery, finely chopped

- 1 medium onion, finely chopped

- 2 medium carrots, finely chopped

- ½ can (from a 7.2 ounces can) tomato sauce

- ½ can (from a 28 ounces can) chopped tomatoes

- ½ box whole wheat spaghetti

- ½ pound ground turkey

- 1 tablespoon olive oil

Directions:

1. Place a pot over medium-high flame. Once the oil is heated, add onion, carrot, and celery and stir. Cook until vegetables are tender.

2. Stir in the ground turkey and garlic and cook until brown, breaking it simultaneously as it cooks.

3. Stir in the tomato sauce and diced tomatoes.

4. While the turkey is cooking, cook spaghetti following the directions on the package. Drain but retain a little of the cooked water, about 3-4 tablespoons.

5. Toss together spaghetti and cannabis tincture and add into the pot. Add a little of the pasta cooked water and mix well.

6. Garnish with parsley and cheese and serve.

Beef and Bean Marijuana Chili

Preparation time: 20 minutes

Cooking time: 4 – 6 hours

Makes: 8 servings

Ingredients:

- 4 ounces dried New Mexico chili flakes (12 – 16 chilies)

- 1 tablespoon ground cumin

- 2 tablespoons salt or to taste

- 2 teaspoons dried oregano

- 2 tablespoons olive oil, divided

- 2 tablespoons canna- oil

- 1 ¾ cups beef stock, divided

- 3 medium onions, chopped

- 10-12 cloves garlic, peeled, minced (about 2 tablespoons)

- 2 tablespoons dark brown sugar

- 2 corn tortillas

- 4 teaspoons freshly ground pepper

- 1 teaspoon cayenne pepper

- 6 pounds boneless, chuck beef, trimmed of fat cubed (¾ inch cubes)

- 1 can (15 ounces) black beans, drained

- 1 can (15 ounces) dark red kidney beans, drained

- 1 cup black coffee

- 4 tablespoons apple cider vinegar

Directions:

1. Place a cast-iron skillet over medium-low flame. When the pan is heated, add chilies into the pan and cook until toasted. You should get an aroma. You need to keep a watch over it as it can get burnt. You need to stir often.

2. Remove the chilies from the pan and place them in a bowl. Pour hot water over the chilies and let it sit for about 30 minutes. Turn the chilies a couple of times.

3. Place the corn tortillas in the skillet one at a time and cook for a minute on either side.

4. Tear up the tortillas and add into a blender. Blend until smooth. Transfer into a bowl.

5. Add chilies (not the soaked water) into a blender. Deseed the chilies if desired and discard the stem.

6. Add ¾ cup beef stock, cumin, cayenne, pepper, canna-oil, oregano, and salt into the blender and blend until smooth.

7. Pour the blended mixture into a slow cooker. Set the slow cooker on high.

8. Add 1 ¼ tablespoon olive oil into the skillet. When the oil is heated, add beef in batches and cook until brown all over. Remove beef with a slotted spoon and place in a bowl.

9. When all of the beef is browned, discard the fat and add the beef into the slow cooker.

10. Add remaining oil into the skillet. When the oil is heated, add onion and garlic and stir-fry until light brown.

11. Transfer the onion mixture into the slow cooker. Also, add the blended tortillas, 1 cup beef stock, vinegar, both the variety of beans, coffee, and brown sugar and stir.

12. Keep the slow cooker covered and set the timer for 4 – 6 hours or until meat is cooked.

13. Ladle into bowls and serve.

Canna-Butter Pan-Seared Steak

Preparation time: 10 minutes

Cooking time: 5 minutes

Makes: 4 – 5 servings

Ingredients:

- 3 pounds ribeye steak, 1 ½ inches thick, at room temperature

- 1 teaspoon salt

- 2 teaspoons canna-butter

- 2 tablespoons olive oil

- Freshly ground black pepper to taste

Directions:

1. Place a large, ovenproof pan over high flame for about 4 – 5 minutes.

2. Brush oil all over the steak and season with salt and pepper.

3. Place steaks in the pan and cook for about half a minute, undisturbed.

4. Spread a little of canna butter over the steaks. Turn the steaks over and cook for half a minute. Spread a little of the canna butter over the steaks. Turn off the heat.

5. Shift the pan into an oven that has been preheated to 350° F and bake for 2 minutes on each side for medium-rare or 3 minutes on each side for medium cooked or 4 minutes for each side for well cooked.

6. Once the steaks are cooked to the desired doneness, take out the steaks from the pan and place on a serving platter. Tent loosely with aluminum foil. Let it sit for 2 minutes.

7. Serve.

Canna-Burgers

Preparation time: 10 minutes

Cooking time: 30 minutes

Makes: 6 – 7 servings

Ingredients:

- 2.2 pounds ground beef or its vegan equivalent

- 0.035 ounce cannabis, first decarbed and then ground into fine powder

- 2 onions, finely chopped

- 2 eggs, beaten

- Salt to taste

- 3 – 4 cloves garlic, peeled, minced

- Pepper to taste

- 1 teaspoon paprika

To serve:

- Tomato slices

- Cucumber slices

- Mayonnaise

- Burger buns

- Lettuce leaves

- Cheese slices etc.

Directions:

1. Add meat, cannabis, onion, eggs, salt, pepper, and cayenne pepper into a bowl.

2. Mix well using your hands. Dip your hands in water before mixing.

3. Divide the mixture into 6 – 7 equal portions and shape into burgers.

4. Place a pan over high flame. Spray some cooking spray in the pan.

5. Cook burgers in the pan for a minute on each side.

6. Now lower the flame to medium. Continue cooking on both the sides until the burgers are cooked to the desired doneness. You can also grill the burgers on a preheated grill.

7. Remove the burgers and place on a plate.

8. If you are using cheese, place the cheese slices on the burgers during the last minute of cooking.

9. Serve with any of the suggested serving options.

Weed-Infused Pulled Pork

Preparation time: 10 minutes

Cooking time: 1 ½ - 2 hours

Makes: 4 – 5 servings

Ingredients:

For pork:

- 2 ½ pounds boneless pork

- 1 ½ tablespoons canna-olive oil or canna-butter

- ½ teaspoon garlic powder

- ½ teaspoon ground cumin

- 6 ounces beer

- 1 ½ tablespoons brown sugar

- ½ tablespoon Himalayan salt

- ½ teaspoon smoked paprika

- Pepper to taste

For sauce:

- 14 tablespoons ketchup

- ¼ cup Dijon mustard

- 1 tablespoon Worcestershire sauce

- 6 tablespoons apple cider

- 2 tablespoons packed dark brown sugar

- 4 – 5 hamburger buns, split

Directions:

1. Trim the fat from the pork and cut into big pieces.

2. Combine canna-oil and spices in a baking dish. Place pork in the dish, and stir pork is until well coated with the mixture.

3. Place the baking dish in an oven that has been preheated to 400° F and bake for about 15 minutes.

4. Pour beer over and around the meat and continue roasting about 2 hours or until the meat is very well cooked and is breaking off.

5. Remove the baking dish from the oven and place it on your countertop.

6. To make sauce: Add ketchup, mustard Worcestershire sauce, apple cider, and dark brown sugar into a bowl and whisk well.

7. When the meat is cool enough to handle, shred the meat with a pair of forks into about 1-½ inch pieces.

8. Add meat into a pot. Add sauce and stir. Heat thoroughly.

9. Toast the buns if desired. Place pulled pork over bottom half of the buns. Cover with top half of the buns and serve.

BBQ Pork Ribs

Preparation time: 30 minutes

Cooking time: About 2 hours

Makes: 4 – 6 servings

Ingredients:

- 2 pounds baby back ribs, remove the tough membrane

- 1 teaspoon onion powder

- 1 teaspoon chili powder

- 1 tablespoon brown sugar

- 1 teaspoon garlic powder

- 1 teaspoon salt

- 1 teaspoon Hungarian paprika powder

- 1 cup cannabis-infused BBQ sauce

Directions:

1. Line a baking sheet with aluminum foil and place meat on it.

2. Add all the spices and salt into a bowl and stir. Sprinkle this mixture all over the meat. Keep the meat along with the baking sheet covered.

3. Place the baking dish in an oven that has been preheated to 350° F and bake for about 1 hour.

4. Uncover and spread cannabis-infused BBQ sauce over the meat. Continue baking for another 30 to 45 minutes.

5. Take out the baking dish from the oven and let the meat sit for 10 minutes.

6. Slice and serve.

Cannabis Garlic and Rosemary Pork Chops

Preparation time: 5 minutes

Cooking time: 20 – 30 minutes

Makes: 2 servings

Ingredients:

- 2 pork chops

- Freshly ground pepper to taste

- 1 clove garlic, minced

- ½ tablespoon canna-oil

- Salt to taste

- ½ tablespoon minced fresh rosemary

- 4 tablespoons canna-butter

Directions:

1. Sprinkle salt and pepper over the pork chops.

2. Add canna-butter, garlic, and rosemary into a bowl. Stir and keep it aside.

3. Place an ovenproof skillet over medium flame. Add canna-oil. When the oil is heated, place pork chops in the pan and cook until golden brown on both the sides. It should take about 4 minutes on each side. Turn off the heat.

4. Brush canna-butter mixture all over the pork chops.

5. Place the skillet into an oven that has been preheated to 350° F and bake for 10 to 13 minutes, depending on how you like it cooked.

6. Divide into 2 plates and serve with the remaining canna-butter mixture.

Jambalaya

Preparation time: 15 minutes

Cooking time: 45 minutes

Makes: 8 – 10 servings

Ingredients:

- 6 tablespoons cannabis olive oil

- 2 medium yellow bell peppers, chopped

- 2 green bell peppers, chopped

- 2 medium onions, chopped

- 2 cans (14.5 ounces each) fire-roasted or regular chopped tomatoes, with its liquid

- 2 packages instant Jambalaya rice mix

- 2 packages (12 ounces each) Andouille sausage, cut into ¼ inch thick slices

- ½ cup chopped fresh parsley (optional)

- 1 1/3 cups water

- 2 pounds large shrimp, peeled, deveined

- 2 teaspoons cayenne pepper or to taste

- Salt to taste

Directions:

1. Place a Dutch oven over low flame. Add canna-oil and let it heat. When the oil is heated, add onion and the bell peppers and cook until slightly tender.

2. Add Jambalaya rice mix, water, and tomatoes with its liquid and stir.

3. Increase the heat to medium and let the mixture come to a boil.

4. Lower the heat once again and cook covered for about 20 minutes. Stir every 7 – 8 minutes.

5. Add shrimp and sausage and keep the pot covered. Continue cooking until the rice is cooked. Stir in cayenne pepper and salt and turn off the heat.

6. Let the pot covered for 10 minutes.

7. Garnish with parsley and serve with rice or pasta.

Baked Shrimp Scampi

Preparation time: 15 minutes

Cooking time: 12 minutes

Makes: 3 servings

Ingredients:

- 1 pound shrimp in shell, peeled, deveined, keep the tails

- 1 tablespoon dry white wine

- 6 tablespoons canna-butter, at room temperature

- 1/8 cup minced shallots

- ½ teaspoon minced fresh rosemary leaves

- ½ teaspoon grated lemon zest

- Yolk of 1 medium egg

- Lemon wedges to serve

- 1 ½ tablespoons canna-oil

- Kosher salt to taste

- 2 teaspoons minced garlic

- 1 ½ tablespoons minced fresh parsley

- 1/8 teaspoon crushed red pepper flakes

- 1 tablespoon fresh lemon juice

- 1/3 cup panko bread crumbs

- Pepper to taste

Directions:

1. Add shrimp into a bowl. Pour wine and oil over it. Sprinkle salt and pepper to taste. Toss well. Set aside for 10 minutes.

2. Add garlic, shallots, herbs, red pepper flakes, butter, lemon juice, lemon zest, panko breadcrumbs, yolk, salt, and pepper into a bowl and mix well.

3. Take a small oval dish of about 9-10 inches. Place shrimp on the bottom of the dish, with the curled tail side facing up. Place them in a single layer.

4. Spoon the breadcrumb mixture evenly over the shrimp.

5. Bake in a preheated oven at 425° F for about 10 -12 minutes or until the mixture is bubbling.

6. For a brown top, broil for a minute.

7. Let it cool for a couple of minutes before serving. Top with lemon wedges and serve.

Grilled Fish Tacos with Ganja Green Salsa

Preparation time: 15 minutes

Cooking time: 15 minutes

Makes: 4 servings

Ingredients:

For salsa:

- 2 large green chilies like Hatch chilies

- 2 tomatillos, remove the husk, halved

- 1 teaspoon minced garlic

- 2 tablespoons canna-oil

- 2 jalapeño peppers

- 1 small onion, quartered

- 2/3 cup minced cilantro

- Lime juice to taste

For fish:

- 2 teaspoons vegetable oil

- 2 pounds mahi-mahi or other firm fish

- Salt to taste

- 2 cups shredded green cabbage

- 2/3 cup crumbled queso fresco

- Pepper to taste

- 2 medium avocadoes, peeled, pitted, diced

- 8 large corn tortillas

Directions:

1. Preheat a grill and place chilies, tomatillos, jalapeños, and onion pieces on the grill and grill until charred. Turn the vegetables on the grill so that it is evenly charred.

2. Remove the vegetables from the grill. Place the chilies and jalapeños in a paper bag and close the bag. Let it sit for 10 minutes.

3. Peel the skin from the chilies and jalapeños and place them in a blender. Also, add cilantro, garlic, lime juice, and canna-oil and blend until pureed.

4. Add salt and pepper and stir. Transfer into a bowl. Cover and set aside.

5. To make fish: Brush oil all over the fish. Sprinkle salt and pepper all over the fish.

6. Set up your grill and preheat it to medium heat. Place fish on the grill and grill for 2 to 3 minutes on each side or until it flakes easily when pierced with a fork.

7. Warm the tortillas following the instructions on the package.

8. Scatter cabbage on the tortillas. Place fish on each tortilla. Drizzle some chili salsa over it. Scatter avocado and cheese on each tortilla and serve.

Weed Fish and Chips

Preparation time: 10 minutes

Cooking time: 10 minutes

Makes: 2 servings

Ingredients:

- 2 large potatoes, peeled, cut into fries
- ½ teaspoon baking powder
- ½ teaspoon pepper
- 1 small egg
- ¾ pound cod fillets
- ½ cup all-purpose flour
- ½ teaspoon salt
- ½ cup canna-milk
- Oil to fry, as required

Directions:

1. Immerse potatoes in a bowl of cold water.

2. Add flour, salt, baking powder, and pepper into a bowl and stir well.

3. Add milk and egg and whisk well. Leave it aside to rest for about 20 minutes.

4. Meanwhile, place a deep fryer pan over medium heat. Pour enough oil to fill the pan (about 2-3 inches). Let the oil heat to 350° F.

5. Add potatoes in batches into the pan and cook until fork tender.

6. Remove potatoes with a slotted spoon and place on a plate lined with paper towels.

7. Dip the fish in the batter, one at a time, and carefully drop it in the hot oil. Cook until golden brown all over.

8. Remove fish with a slotted spoon and place on a plate lined with paper towels.

9. Once you are done with the frying of fish, add the potatoes back into the hot oil and cook for a couple of minutes until crisp.

10. Serve fish with chips.

Sativa Shrimp Creole

Preparation time: 15 minutes

Cooking time: 30 minutes

Makes: 2 servings

Ingredients:

- ½ tablespoon unsalted butter
- 1 tablespoon all-purpose flour
- ¼ green bell pepper, diced
- 1 teaspoon minced garlic
- ¼ small onion, diced
- ½ stalk celery, finely diced
- ½ can (from a 15 ounces can) crushed tomatoes with its juices
- 2 small bay leaves
- 1 tablespoon chopped fresh Italian parsley or 1 teaspoon dried parsley
- 0.004 ounce decarbed kief or finely ground hash
- Cooked rice to serve
- ½ tablespoon olive oil
- ½ cup stock
- A pinch cayenne pepper or to taste
- Salt to taste
- ½ pound medium shrimp, peeled

Directions:

1. Place a skillet over medium flame. Add butter and olive oil and let the butter melt.

2. Add flour and stir continuously until roux is formed and is light brown in color.

3. Stir in celery, bell pepper, and onion and keep stirring until the vegetables are slightly tender.

4. Add garlic and cook for a few seconds until you get a nice aroma.

5. Add tomatoes with its juices, bay leaves, parsley, kief, stock, cayenne pepper, and salt, then stir.

6. Bring it to a boil over high heat. Now lower the heat to low heat and cook for about 10 minutes.

7. Stir in shrimp and cook until pink.

8. Divide rice into plates. Divide shrimp creole and spoon over the rice. Serve immediately.

Butter Garlic Shrimp

Preparation time: 10 minutes

Cooking time: 15 minutes

Makes: 8 servings

Ingredients:

- 2 pounds shrimp, peeled, deveined

- 1 cup canna-butter

- ½ teaspoon red pepper flakes

- 2/3 cup chopped parsley

- Angel hair pasta, to serve (optional)

- 4 tablespoons canna-olive oil

- 12 cloves garlic, peeled, minced

- 6 tablespoons lemon juice

- Salt to taste

Directions:

1. Place a large skillet over medium flame. Add canna-olive oil and let it heat. Add shrimp into the pan and spread it evenly. Do not stir for 3 – 4 minutes.

2. Sprinkle salt over it and stir. When shrimp begins to become light pink, stir in garlic and red pepper flakes.

3. Sauté for a couple of minutes. Stir in lemon juice, ½ cup canna-butter, and 1/3-cup parsley. When butter melts, lower the heat and add remaining butter.

4. Let it simmer until slightly thick.

5. Remove shrimp with a slotted spoon and add into a bowl. Add 1/3-cup parsley and toss well.

6. Let the sauce simmer for 3 – 4 minutes. If the consistency of the sauce is thick, add water to dilute, using 1 teaspoon at a time, stirring each time.

7. Turn off the heat. Add salt to taste and stir.

8. To serve as main course, serve over angel hair pasta. Drizzle sauce on top and serve. You can also serve it with steak.

9. To serve as appetizer, serve shrimp on a serving platter with sauce in a bowl as a dip.

Cannabis-Infused Pasta with Clams and Green Chiles

Preparation time: 15 minutes

Cooking time: 30 minutes

Makes: 2 – 3 servings

Ingredients:

- ¼ cup + ½ tablespoon extra-virgin olive oil

- ¼ cup lightly packed parsley leaves

- 1 small poblano pepper, discard stem, deseeded, chopped

- 1 ounce shishito peppers, discard stems, deseeded, chopped

- Kosher salt to taste

- 1 shallot, thinly sliced

- 1 small shallot, minced

- 1 tablespoon drained capers

- ½ cup dry white wine

- 21 mixed clams like Manila, littleneck and razor, scrubbed

- 1 ½ tablespoons unsalted butter

- 2 tablespoons crème fraiche

- Wasabi caviar, to garnish

- A large handful mint leaves + extra to garnish

- 2 tablespoons snipped chives

- 1 Cubanelle pepper, discard stem, deseeded, chopped

- Pepper to taste

- 1 clove garlic, minced

- 1 clove garlic, crushed

- 1 teaspoon whole peppercorns

- 1 teaspoon fennel seeds

- 1 teaspoon coriander seeds

- 1 teaspoon mustard seeds

- 1 cup bottled clam juice

- 6 ounces pipe rigate or mezze rigatoni pasta

- ½ tablespoon canna-butter

Directions:

1. Add ¼ cup oil, parsley, half the mint leaves, and chives into a blender and blend until smooth.

2. Place a fine wire mesh strainer over a bowl. Pass the blended mixture through the strainer. Press to remove as much liquid as possible. Throw away the solids.

3. Place a cast-iron skillet over medium-high flame. Add ½ tablespoon oil and let it heat well, almost up to smoking point.

4. Place all the varieties of chilies in the pan. Season with salt and pepper and cook until charred with few blisters on the peppers. Stir occasionally.

5. Stir in the minced shallot and minced garlic. Add capers and cook on low for a couple of minutes and turn off the heat.

6. After about 10 minutes, add the strained oil into the pan and stir.

7. Place a pot over medium flame. Add whole spices into the pan and cook for 2 – 3 minutes, stirring frequently, until you get a nice aroma in the air.

8. Now stir in the sliced shallot and crushed garlic. Pour clam juice and stir.

9. When the juice comes to a boil, add clams, and cover the pot. Increase the heat to high heat and cook for about 8 – 9 minutes. By now, the clams should have opened up. Discard any clams that have not opened up.

10. Remove clams with a pair of tongs and place them on a baking sheet. When it cools a bit, take out the meat from the shells and throw off the shells.

11. Place a strainer over a bowl and pass the cooked liquid through the strainer. Use only the liquid and not the solids; they can be thrown away.

12. In the meantime, cook pasta, following the directions on the package. Drain and set aside.

13. Wipe the pot in which you have cooked pasta. Place the pot over medium-high flame.

14. Add butter. When butter melts, add clams, pasta, ¼ cup clam cooked liquid, chili mixture, and crème fraiche and stir. Heat thoroughly.

15. Add remaining mint leaves, canna-butter, lime juice, salt, and pepper and mix well. Turn off the heat.

16. Garnish with mint leaves and wasabi caviar and serve.

Weed Grilled Cheese

Preparation time: 5 minutes

Cooking time: 6 minutes

Makes: 2 servings

Ingredients:

- 2 slices cheddar cheese

- 4 teaspoons butter

- 4 teaspoons canna-butter

- 4 slices bread

Directions:

1. Apply a teaspoon of butter on one side of each of the bread slices.

2. Apply a teaspoon of canna-butter on the other side of each of the bread slices.

3. Lay the cheese slice over the canna-butter side, on 2 slices of bread. Cover with the remaining 2 slices of bread, with the canna-butter side facing down.

4. Place a nonstick pan over medium flame. When the pan is heated, place the sandwich in the pan and cook until the bottom side is golden brown. Turn the sandwich over and cook the other side until golden brown. Remove the sandwich and cut into the desired shape.

5. Cook the other sandwich similarly.

6. This tastes great with tomato soup.

Smoked Mac 'n' Cheese

Preparation time: 10 minutes

Cooking time: 50 – 60 minutes

Makes: 2 – 3 servings

Ingredients:

- ¼ cup unsalted butter

- ¼ cup cold canna- butter

- ½ tablespoon melted canna-butter

- 2 cups milk

- 1 teaspoon salt or to taste

- ½ teaspoon pepper or to taste

- ½ cup grated, smoked mozzarella cheese

- ½ cup grated parmesan cheese, divided

- ½ cup shredded cheddar cheese

- ½ cup shredded American of Swiss cheese

- ½ pound penne pasta

- ½ cup flour

- 1/8 teaspoon cayenne pepper or to taste

- 2 tablespoons breadcrumbs

Directions:

1. Cook pasta, following the directions on the package.

2. Place a skillet over medium flame. Add butter and cold canna-butter. When butter melts, stir in flour and cook for 2-3 minutes until roux is formed and cooked to the desired color.

3. Meanwhile, heat milk in a saucepan. When it nearly begins to boil, but not boiling, pour milk into the skillet, stirring constantly. Stir in salt, pepper, and cayenne pepper.

4. Cook until thick. Keep stirring until the mixture begins to boil. Turn off the heat. Add pasta, smoked mozzarella cheese, half the parmesan cheese, cheddar cheese, and American cheese. Mix well.

5. Spray a baking dish with some cooking spray.

6. Transfer the mixture into the baking dish.

7. Add breadcrumbs into a bowl. Add melted butter and mix well. Scatter this mixture over the pasta in the baking dish.

8. Place the baking dish in a preheated oven and bake at 400° F until golden brown on top.

9. Serve.

Marijuana Pizzadillas

Preparation time: 5 minutes

Cooking time: 1 minute

Makes: 2 servings

Ingredients:

- 4 small flour tortillas

- 2/3 cup pizza sauce

- ½ cup grated mozzarella cheese

- 1 teaspoon canna-oil

- ¼ - ½ cup pepperoni or cooked sausage (optional)

- ½ cup finely chopped vegetables of your choice (optional)

Directions:

1. Combine pizza sauce and canna-oil in a bowl.

2. Take 2 paper plates and place a tortilla on each. Scatter 2 tablespoons cheese on each tortilla.

3. Drop the pizza sauce in small amounts all over the tortillas. Scatter pepperoni and vegetables if using. Scatter remaining cheese all over the vegetables.

4. Cover with the remaining tortillas. Place the Pizzadillas in the microwave, one at a time, and cook for 1 minute each.

5. Remove from the microwave and let it cool for about a minute.

6. Cut into wedges and serve.

Cannabis-Infused French Bread Pizza

Preparation time: 5 minutes

Cooking time: 10 minutes

Makes: 2 servings

Ingredients:

For sauce:

- 1 ounce tomato paste
- ½ tablespoon canna-oil
- ¼ teaspoon minced garlic
- Crushed red pepper flakes to taste
- 1 ounce water
- ½ tablespoon olive oil
- ½ teaspoon Italian seasoning blend
- ¼ teaspoon balsamic vinegar

For pizza:

- 6 tablespoons shredded mozzarella cheese
- 1 small, soft loaf French or Italian bread, halved lengthwise
- 2 tablespoons grated parmesan cheese
- Toppings of your choice like onion, pepperoni, mushrooms, etc.

Directions:

1. To make pizza sauce: Combine tomato paste, canna-oil, garlic, red pepper flakes, water, olive oil, Italian seasoning, and vinegar in a bowl.

2. Place the bread halves on a plate, with the cut side facing up. Apply half the sauce on each bread. Scatter mozzarella on the bread slices. Next, scatter Parmesan, followed by any toppings.

3. Place rack in the center of the oven and preheat it to 400° F.

4. Place the bread pizzas on the rack and bake until cheese melts and is brown at a few spots.

5. Serve hot.

Fettuccine Alfredo Pasta

Preparation time: 15 minutes

Cooking time: 15 minutes

Makes: 3 servings

Ingredients:

- 12 ounces dry fettuccine pasta

- ¾ cup heavy cream

- ¼ cup grated parmesan cheese

- 1/3 heaping cup grated Romano cheese

- Garlic salt to taste

- Salt to taste (optional)

- ½ cup canna-butter

- Pepper to taste

Directions:

1. Cook fettuccini pasta in salted water, following the directions on the package.

2. Place a saucepan over low flame. Add canna-butter and let it melt.

3. Stir in cheese, salt, garlic salt, and pepper. Stir often until cheese melts.

4. Add cooked pasta and toss well.

5. Serve.

Side Dish

Mini Cannabis Green Bean Casserole

Preparation time: 15 minutes

Cooking time: 30 minutes

Makes: 8 servings

Ingredients:

- 6 cups fresh green beans, trimmed, chopped into 2 inch pieces

- 4 tablespoons butter

- 6 tablespoons canna-butter

- 1 large onion, chopped

- 2 cans (12.8 ounces each) cream of mushroom soup

- Salt to taste

- 2 cups sliced mushrooms

- 3 cups French fried onions, divided

- Pepper to taste

Directions:

1. Place a pot of water over high heat. When the water begins to boil, add green beans and cook for a few minutes until crisp as well as tender.

2. Drain in a colander.

3. In the meantime, place a large skillet over medium heat. Add canna-butter.

4. When the butter melts, add onion and mushrooms and sauté until tender. Turn off the heat.

5. Add beans, half the French fried onions, pepper, salt, and cream of mushroom soup and mix well.

6. To prepare ramekins and oven: Grease 8 ramekins with butter. You need to preheat the oven to 375° F.

7. Spoon the green bean mixture into the prepared ramekins. Scatter remaining French fried onions on top.

8. Bake until the top is golden brown.

Mashed Cannabis Cauliflower with Parmesan Cheese

Preparation time: 10 minutes

Cooking time: 10 minutes

Makes: 2 – 3 servings

Ingredients:

- 1 ¼ pounds cauliflower, cut into florets

- 3 tablespoons heavy whipping cream

- ¼ teaspoon pepper

- ½ cup shredded parmesan cheese, divided

- ½ tablespoon canna-butter

- 1 tablespoon minced parsley

- Salt to taste

Directions:

1. Place a pot half-filled with water over high heat. When the water begins to boil, add cauliflower florets and cook for a few minutes until tender.

2. Turn off the heat and keep the pot covered for about 15 minutes. Drain in a colander.

3. Transfer the cauliflower into a bowl and mash it with a potato masher. Add half the cheese, canna-butter, cream, salt, and pepper.

4. Garnish with rest of the cheese and parsley.

Cannabis Mashed Potatoes

Preparation time: 15 minutes

Cooking time: 20 minutes

Makes: 2 – 3 servings

Ingredients:

- 1 ¼ pounds russet potatoes, peeled, cut into 1 ½ inch cubes

- 2 teaspoons canna-butter

- 1 ounce cream cheese

- Kosher salt to taste

- Freshly ground pepper to taste

- 1 tablespoon butter, at room temperature

- Chopped chives to garnish

Directions:

1. Add water to a pot and place the pot over medium heat.

2. Add potatoes and about a teaspoon of salt into it. Cook until the potatoes are soft.

3. Drain the potatoes and discard the water. Add the potatoes, canna-butter, butter, cream cheese, salt, and pepper into a bowl. Mash with an electric hand mixer until smooth, or the texture you prefer is achieved.

4. Garnish with chives and serve.

Caramelized Brussels Sprouts

Preparation time: 5 minutes

Cooking time: 20 minutes

Makes: 2 – 3 servings (4 Brussels sprouts per serving)

Ingredients:

- 1 slice uncured bacon, cut into pieces
- ½ pound fresh Brussels sprouts, halved
- 2 cloves garlic, minced
- 1 tablespoon light agave nectar or maple syrup
- 1 tablespoon canna-oil
- ½ small onion, thinly sliced
- Pepper to taste
- 2 tablespoons balsamic vinegar or apple cider vinegar
- Salt to taste

Directions:

1. Place a pan over medium flame. Add bacon and cook until slightly crisp. Remove bacon with a slotted spoon and place it on a plate lined with paper towels.

2. Add canna-oil into the pan. Once oil is heated, add onion and cook until light golden brown.

3. Stir in Brussels sprouts and garlic and until brown, stirring occasionally.

4. Stir in the vinegar, salt, pepper, and agave nectar. Cook until dry.

5. Transfer into a bowl and serve topped with bacon.

Canna-Butter Sautéed Mushrooms

Preparation time: 5 minutes

Cooking time: 5 – 7 minutes

Makes: 2 servings

Ingredients:

- 1 – 2 tablespoons canna-butter

- ½ pound mushrooms, halved or quartered or whole, according to the size

Directions:

1. Place a pan over medium flame. Add canna-butter. When butter melts, add mushrooms and sauté until tender.

2. Serve immediately.

Broccoli Cheddar Cannabis Casserole

Preparation time: 10 minutes

Cooking time: 45 minutes

Makes: 4 servings

Ingredients:

- 5 tablespoons canna-butter, melted

- 4 cups crushed Ritz crackers or any other crackers

- 2 cups mayonnaise

- 2 cups grated sharp cheddar cheese

- 2 ½ pounds broccoli, cut into bite-size florets

- 4 eggs, lightly beaten

- 2 cups condensed cream of mushroom soup

Directions:

1. Place a pot of water over high heat. When the water begins to boil, add broccoli and cook for a few minutes until crisp as well as tender. They will turn bright green in color.

2. Drain in a colander. Immediately immerse broccoli in a bowl of ice water. Drain after 5 minutes in a colander.

3. Add broccoli, eggs, mushroom soup, cheddar cheese, and mayonnaise into a bowl and mix until broccoli is well coated with the mixture.

4. You need to prepare a baking dish and oven: For the oven, preheat it to 375° F. Spray a baking dish with cooking spray.

5. Add the broccoli mixture into the casserole dish

6. Pour melted canna-butter all over the broccoli mixture and swirl the pan to spread the butter all over.

7. Place the casserole dish in the oven and bake for about 30 – 40 minutes or until light golden brown on top.

8. Serve.

Cannabis-Infused Radical Ratatouille

Preparation time: 15 minutes

Cooking time: 20 minutes

Makes: 3 servings

Ingredients:

- 1 ½ tablespoons olive oil

- ½ tablespoon minced garlic

- 1 small zucchini, 1 inch dice

- ½ bell pepper or any color, cut into 1 inch squares

- 1 cup cubed eggplant, 1 inch cubes

- ½ cup cubed yellow summer squash

- ¾ cup diced tomatoes

- ½ large onion, cut into 1 inch dice

- ¼ teaspoon dried thyme

- ½ teaspoon pepper

- ½ teaspoon dried oregano

- Salt to taste

- 1 tablespoon canna-oil

- A handful fresh cilantro, chopped

Directions:

1. Place a cast-iron skillet over medium-high flame. Add ¾ tablespoon oil and let it heat.

2. Add onion and cook until slightly tender. Add garlic and cook for a few seconds until fragrant.

3. Stir in remaining oil and eggplant and cook for 4 – 5 minutes. Stir in squash and zucchini. Stir often and cook until slightly tender.

4. Stir in tomatoes, dried herbs, pepper, and salt. Cook until vegetables are tender.

5. Add canna-oil and basil and mix well.

6. Serve.

Almond, Orange, and Cucumber Stuffed Avocado

Preparation time: 15 minutes

Cooking time: 10 minutes

Makes: 4 servings

Ingredients:

- 6 tablespoons balsamic vinegar

- 4 small Clementine oranges, peeled, separated into segments

- A handful fresh cilantro, chopped

- ½ teaspoon fresh lemon juice

- Salt to taste

- 2 large avocadoes, halved, pitted

- 2 small Persian cucumbers, peeled, diced

- 3 tablespoons sliced, toasted almonds

- Pepper to taste

- 1 ½ tablespoons canna-olive oil

Directions:

1. Pour vinegar into a small saucepan. Place the saucepan over medium flame. Cook until it is reduced to 3 tablespoons.

2. Scoop out the avocado and cut into small cubes. Leave the shells of the avocado with a little of the flesh on it so that it remains stable when you fill it up.

3. Add avocado into a bowl along with rest of the ingredients. Mix well and fill this mixture into the avocado halves.

4. Trickle reduced vinegar on top and serve.

Weed Bread

Preparation time: 15 – 20 minutes + rising time

Cooking time: 40 – 50 minutes

Makes: 2 loaves or 1 large loaf

Ingredients:

- 0.35 ounce marijuana flowers, deseeded, de-stemmed

- Flaxseed meal or cornmeal, to dust

- ½ teaspoon active dry yeast

- 6 cups all-purpose flour or mixture of 4 cups flour and 2 cups whole wheat flour

- 2 ½ teaspoons salt

- 3 ½ cups lukewarm water (110°F)

Directions:

1. Decarb the marijuana and grind it coarsely.

2. Combine marijuana flowers, yeast, flour, and salt in a mixing bowl. Pour lukewarm water and mix to form sticky dough.

3. Keep the bowl covered with cling wrap and place it in a warm area, say on top of your refrigerator, for 12-16 hours or until the dough doubles in size and tiny bubbles should be visible on the top of the dough.

4. Dust your countertop as well as your hands with cornmeal or flaxseed meal.

5. Place the dough on your countertop. Divide the dough into 2 equal portions. You can also make one large loaf. Fold each portion of the dough a few times to form into a ball. Dredge the dough in some more cornmeal.

6. Place dough on a cotton cloth, with its seam side facing down. Sprinkle some cornstarch on top of the dough.

7. Wrap the dough lightly with the cloth and set aside for 2 hours to rise.

8. You need to preheat the oven to 450° F for 30 minutes before baking. Place a large metal pan or Dutch oven in the oven while preheating.

9. Remove the dough from the cloth and place in the heated loaf pan carefully with its seam side facing up.

10. Cover with a lid of a dutch oven or aluminum foil, place it in the oven, and bake for 40-50 minutes.

11. Uncover and continue baking for another 8-10 minutes. This is done to get a crust on top.

12. Switch off the oven and let the bread remain in the oven for 5 minutes. When you tap the bread, you should hear a hollow sound if the bread is ready.

13. Carefully remove the bread from the pan and cool on a wire rack completely.

14. Slice and serve. Store leftovers in an airtight container.

Cannabis Gravy

Preparation time: 5 minutes

Cooking time: 10 minutes

Makes: 20 – 24 servings

Ingredients:

- 1 cup canna-butter

- Salt to taste

- 4 onions, thinly sliced

- 1 cup balsamic vinegar

- Chicken stock or turkey stock, as required

- Pepper to taste

- ¼ chopped cup fresh sage,

- ¼ chopped cup fresh rosemary

- 2/3 cup flour

Directions:

1. Place a large skillet over low flame. Add canna-butter.

2. When the butter just melts, add onions and sauté until translucent. Make sure that the flame is low all the time.

3. Add rosemary and sage and cook for another 10 minutes.

4. Add flour and sauté for about 60 to 90 seconds, stirring constantly.

5. Pour the stock, stirring simultaneously. Cook until the gravy thickens.

6. Add vinegar and simmer for just 15 minutes and not longer.

7. Add salt and pepper to taste.

Weed French Fries

Preparation time: 15 minutes

Cooking time: 20 minutes

Makes: 2 servings

Ingredients:

- 1/3 cup canna- extra-virgin olive oil

- 1 ½ tablespoons salt

- 2 large potatoes, peeled, cut into fries

Directions:

1. Place the potatoes on a baking sheet.

2. Sprinkle salt and oil all over them. Toss well. Spread it evenly all over the baking sheet without overlapping.

3. Place the baking sheet in an oven that has been preheated to 400° F and bake for 20 to 30 minutes, depending on the way you like it cooked.

4. Serve hot.

CHAPTER 6

DESSERT RECIPES

Chocolate Pudding

Preparation time: 10 minutes

Cooking time: 15 minutes

Makes: 4 – 6 servings

Ingredients:

- 2/3 cup sugar

- 1/3 cup cocoa powder

- 2 cups milk

- 2 tablespoons grated chocolate

- 2 tablespoons canna-butter

- 3 tablespoons cornstarch

- ½ teaspoon vanilla extract

Directions:

1. Add cocoa, sugar, cornstarch, and salt into a small pot and whisk well.

2. Add milk and whisk until smooth.

3. Place the pot over medium flame and stir constantly until it comes to a boil. After about 30 seconds, turn off the heat.

4. Stir in vanilla and canna butter. Pour the mixture into pudding cups or bowls. Let it cool completely.

5. Chill until use. Garnish with grated chocolate and serve.

French Toast Cupcakes

Preparation time: 15 minutes

Cooking time: 25 minutes

Makes: 6 servings

Ingredients:

For cupcakes:

- ¾ cup all-purpose flour

- ¾ teaspoon baking powder

- ¼ teaspoon ground allspice

- ¼ teaspoon salt

- ½ cup sugar

- ½ teaspoon ground cinnamon

- 1/8 teaspoon freshly grated nutmeg

- ¼ teaspoon ground allspice

- ¼ cup canna-butter

- 1 large egg

- 2 slices cooked bacon, cut each into 3 pieces

- ¼ cup sour cream

For topping:

- 2 tablespoons all-purpose flour

- 2 tablespoons chopped pecans

- 1 ¼ tablespoons butter, cubed, chilled

- 2 tablespoons sugar

- ¼ teaspoon ground cinnamon

Directions:

1. To make topping: Add flour, butter, pecans, sugar, and cinnamon into a bowl and mix until the butter is pea-size. Keep the bowl in the refrigerator, covered.

2. To prepare oven and muffin pan: Take a 6 counts muffin pan and place disposable paper liners in them. Place rack in the center of the oven and preheat the oven to 350° F.

3. Combine flour, baking powder, salt, spices, and sugar in a bowl.

4. Add sour cream, canna-butter, maple extract, and egg in a bowl and beat with an electric hand mixer set on medium speed until creamy.

5. Lower the speed and add the mixture of dry ingredients. Continue beating until just incorporated, making sure not to overbeat.

6. Pour batter into the prepared muffin tins, up to ¾ the muffin cups.

7. Divide equally the topping and scatter it over the batter in the muffin cups.

8. Place the muffin pan in the oven and bake for about 25 minutes. When the muffins are ready, if you pierce a toothpick in the middle of the cupcake, it should come out without any particles stuck on it.

9. Take out the muffin pan from the oven and let it remain on your countertop for 15 minutes.

10. Remove the muffins from the pan and place it on a cooling rack. Insert a piece of bacon in each muffin and serve.

11. Store leftovers in an airtight container in the refrigerator.

Key Lime Kickers

Preparation time: 5 minutes

Cooking time: 20 minutes

Makes: 12 servings

Ingredients:

- 3 tablespoons heavy cream

- 2 tablespoons weed sugar

- 4-5 drops key lime oil

- 1 tablespoon unsalted butter

- 5 ounces white chocolate, coarsely chopped

- Graham cracker crumbs to coat

Directions:

1. To make ganache, set up double boiler: Take 2 pots of nearly the same (but not same) sizes such that the smaller one fits inside, the larger pot.; the smaller pot should not touch the bottom of the bigger pot. It should fit well inside it.

2. Pour enough water into the larger pot such that it is 1/3 full. The water should not touch the smaller pot. Place the bigger bowl over medium flame. Let the water come to a boil.

3. Add cream into the smaller pot. Place the smaller pot inside the bigger pot.

4. Lower heat to low heat and let the water simmer. Add corn syrup, butter, and weed sugar and stir.

5. Stir in white chocolate. Stir occasionally. When the mixture is smooth, remove the bowl from the double boiler.

6. Take a bowl to make the dessert and weigh the bowl. Note down the weight.

7. Pat the bowl dry from the sides and bottom of the pan. Pour into the weighed bowl.

8. Place this bowl in the freezer until set, but not hard. You should be able to shape the ganache. Now weigh the bowl once again.

9. Use this formula: Weight of the bowl with ganache – (minus) weight of the bowl without ganache divided by 24. This formula is to get the weight of a truffle.

10. Add cracker crumbs into a shallow bowl.

11. Scoop out ganache (equal to the weight got from the formula) and place it on a baking sheet lined with parchment paper.

12. Dredge the ganache in cracker crumbs and place on another baking sheet lined with parchment paper.

13. Cover the baking sheet and chill until use. It can last for 4 – 5 weeks.

Brownies

Preparation time: 10 – 12 minutes

Cooking time: 20 minutes

Makes: 9 servings

Ingredients:

- 2 tablespoons salted butter

- 6 tablespoons canna-butter

- 2 ounces unsweetened chocolate

- 5 ounces semi-sweet chocolate

- 3 tablespoons unsweetened cocoa powder

- 1 ¼ cups sugar

- ½ tablespoons pure vanilla extract

- 3 large eggs

- ½ teaspoon salt

- 1 cup all-purpose flour

For drizzle:

- 1/8 teaspoon vegetable shortening

- ¼ cup white chocolate chips

Directions:

1. To prepare baking dish and oven: Spray a square baking dish (8 – 9 inches) with cooking spray and set it aside. Place rack in the center of the oven and make sure that the oven is preheated to 350° F.

2. Place a saucepan over low flame. Add canna-butter and let it melt. Add both the chocolates and keep stirring until chocolates melt.

3. Add cocoa powder and whisk well. Turn off the heat.

4. Add egg in a bowl and whisk. Add sugar, vanilla, and salt and whisk well. Pour the melted chocolate mixture and keep whisking until well incorporated.

5. Add flour and fold gently. Spoon the batter into the baking dish. Place the baking dish in the oven and bake until firm on top. It should take about 20 minutes. When ready, a toothpick, when pierced in the middle of the brownie, comes out without any particles stuck on it.

6. Let the baking dish cool on your countertop.

7. Cut into 9 equal pieces.

8. To make drizzle: Melt chocolate and shortening in a double boiler. The method is given in the previous recipe. Keep stirring until chocolate melts.

9. Drizzle the chocolate over the brownies.

Canna Chocolate Dipped Strawberries

Preparation time: 5 minutes

Cooking time: 2 minutes

Makes: 6 servings

Ingredients:

- 1 tablespoon canna-coconut oil

- 6 strawberries with stem

- ¾ cup chocolate chips

Directions:

1. Add chocolate chips and coconut oil into a microwave-safe bowl. Cook on high for about a minute. Stir every 12 – 15 seconds until chocolate melts completely.

2. Line a baking sheet with parchment paper. Hold the strawberries with its stem and dip it in the chocolate. Lift it and place it on the baking sheet. Once chocolate sets, it is ready to serve.

Pineapple Upside-Down Cake

Preparation time: 15 minutes

Cooking time: 50 – 60 minutes

Makes: 18 – 20 servings

Ingredients:

- 12 – 14 canned pineapple slices, drained

- 4 cups granulated sugar

- 4 cups cake flour

- 2 teaspoons salt

- 3 tablespoons dark rum

- 4 eggs

- 1 ½ cups canna-butter, at room temperature

- 1 ¼ cups firmly packed light brown sugar

- 2 ¼ teaspoons baking powder

- 1 ½ cups milk

- 3 teaspoons vanilla extract

- 12 – 14 maraschino cherries

Directions:

1. To prepare oven and baking dish: Place rack in the lower third position in the oven and make sure that the oven is preheated to 350° F.

2. Grease 2 pie pans (9 inches) with some oil or butter.

3. Divide equally the pineapple slices and lay them on the bottom of the pan, next to each other, without overlapping.

4. Add ¾ cup canna-butter, 1-cup light brown sugar, and 1 cup granulated sugar into a saucepan.

5. Place the saucepan over medium flame. Keep stirring until butter melts. Remove the pan off the heat and keep stirring until sugar is dissolved completely.

6. Divide the sugar solution equally and pour it all over the pineapple slices.

7. Sift cake flour, salt, and baking powder in a bowl.

8. Combine rum, milk, ¼ cup brown sugar, and vanilla in a bowl.

9. Add remaining canna-butter and remaining sugar into a mixing bowl. Beat with an electric hand mixer, set on high speed until light and creamy.

10. Beat in the eggs, one egg each time. Beat well each time. Add vanilla extract.

11. Set the mixer on low speed and add the mixture of flour and mixture of milk, a little at a time and beat until just combined, making sure not to over-beat.

12. Divide equally the batter among the pie pans, over the, over the pineapple slices.

13. Bake the cakes in a preheated oven at for about 50 – 60 minutes or until a toothpick, when pierced in the middle of the cake, comes out without any particles stuck on it.

14. Switch off the oven and let the pie pans remain in the oven for 10 minutes.

15. Remove the pie pans from oven and let them cool for 15 minutes. Run a knife around the edges of the cake and invert on to 2 plates.

16. Cut each into 9 – 10 slices. Place a cherry on each piece and serve.

17. Store leftovers in an airtight container in the refrigerator. Remove from the refrigerator an hour before serving.

Macadamia & White Chocolate Cookies

Preparation time: 15 minutes

Cooking time: 10 minutes

Makes: 12 – 15 servings

Ingredients:

- ½ cup canna-butter, at room temperature

- ½ cup chopped macadamia nuts

- 6 tablespoons packed, light brown sugar

- ½ cup chopped white chocolate

- 1 ¼ cups flour

- ¼ cup cane sugar

- ½ teaspoon baking soda

- ¼ teaspoon vanilla extract

- ¼ teaspoon almond extract

- ¼ teaspoon salt

- 1 egg

Directions:

1. Add canna-butter, cane sugar, and brown sugar into a mixing bowl and mix with an electric hand blender until creamy.

2. Add egg and beat well. Add vanilla extract and almond extract and beat well.

3. Add flour, baking soda, and salt into a bowl and stir until well combined. You can also sift them together.

4. Add the flour mixture, about 2 tablespoons at a time, and mix well each time.

5. When all of the flour mixture is added, stir in the nuts and chocolate.

6. You need a baking sheet without any grease on it.

7. Scoop out the mixture and drop them on the baking sheet. You should have about 12 – 15 cookies.

8. Place the baking sheet in an oven that has been preheated to 350° F and bake for 10 minutes or until golden brown around the edges.

9. Let the cookies cool on the baking sheet.

10. Transfer into an airtight container and store at room temperature. It should last for 7 – 10 days.

Marijuana Chocolate Chip Cookies

Preparation time: 10 minutes

Cooking time: 10 – 12 minutes

Makes: 25 – 30 servings

Ingredients:

- 4 ¾ cups flour

- 2 teaspoons salt

- 2 ounces butter

- 2 big cups brown sugar

- 4 large eggs

- 2 teaspoons baking soda

- 12 ounces canna-butter

- 1 ½ cups sugar

- 2 teaspoons vanilla extract

- 3 ½ cups chocolate chips

Directions:

1. Sift flour, salt, and baking soda into a bowl.

2. Add canna-butter, brown sugar, vanilla, and sugar into a mixing bowl. Beat with an electric hand mixer set on high speed. Keep beating until fluffy.

3. Beat in the eggs, one at a time, and beat well each time.

4. Add the mixture of dry ingredients and mix well.

5. Add chocolate chips and mix well.

6. You need 1 – 2 baking sheets lined with parchment paper.

7. Scoop out the mixture and drop them on the baking sheet. You should have about 25 – 30 cookies.

8. Place the baking sheet in an oven that has been preheated to 375° F and bake for 10 minutes or until golden brown around the edges.

9. Let the cookies cool on the baking sheet.

10. Transfer into an airtight container and store at room temperature. It should last for 7 – 10 days.

Weed Chocolate Bars

Preparation time: 5 minutes

Cooking time: 10 – 15 minutes

Makes: 4 servings

Ingredients:

- ½ cup canna-butter

- 6 tablespoons powdered sugar

- 6 tablespoons unsweetened cocoa

Directions:

1. Take 2 pots of nearly the same (but not same) sizes such that the smaller one fits inside, the larger pot., The smaller pot should not touch the bottom of the bigger pot. It should fit well inside it.

2. Pour enough water into the larger pot such that it is 1/3 full. The water should not touch the smaller pot. Place the bigger bowl over medium flame. Let the water come to a boil.

3. Add butter into the smaller pot. Place the smaller pot inside the bigger pot.

4. Lower heat to low heat and let the water simmer. Once butter melts, remove the smaller pot from the double boiler.

5. Sift together cocoa and powdered sugar and add into the melted butter. Stir until well incorporated. Take a chocolate bar mold and pour the chocolate mixture into the mold.

6. Once completely cooled, place in the refrigerator for a couple of hours or until set.

7. Unmold and serve.

Vegan Pumpkin Spice Ice Cream

Preparation time: 5 minutes

Cooking time: 0 minutes

Makes: 10 – 12 servings

Ingredients:

- 1 – 2 teaspoons pumpkin pie spice

- 4 cans full fat coconut milk

- 1 teaspoon vanilla extract

- A large pinch sea salt

- 2 – 4 tablespoons maple syrup

- 2 – 4 tablespoons canna-coconut oil, melted

Directions:

1. Shake the cans of coconut milk well and pour into ice cube trays. Freeze until use. If you do not have sufficient ice trays, pour into a pan lined with parchment paper and freeze until firm. You can then chop them into chunks and use as required.

2. Add coconut milk ice cubes, pumpkin pie spice, vanilla, salt, maple syrup, and canna-coconut oil into the food processor bowl. Process until well combined.

3. For soft-serve consistency, you can serve right away.

4. For firm ice cream, transfer into a freezable bowl and freeze until firm.

Weed Vanilla Ice Cream

Preparation time: 30 minutes

Cooking time: 5 minutes

Makes: 8 servings

Ingredients:

- 1 ½ cups white sugar

- 4 ½ cups canna-milk

- 2 cups heavy whipping cream

- 4 teaspoons vanilla extract

Directions:

1. Combine sugar, canna-milk, cream, and vanilla in a saucepan.

2. Place the saucepan over low flame and stir frequently until sugar dissolves. Turn off the heat and pour into a bowl.

3. Add vanilla extract and stir. Cool completely. Cover the bowl with cling wrap and chill for 7 – 8 hours.

4. Add the mixture into an ice cream maker. Follow the instructions of the manufacturer and make the ice cream.

5. You can serve right out of the ice cream maker for a soft-serve consistency. Else transfer into a freezer-safe container and freeze until firm.

6. If you do not have an ice cream maker, pour into a freezer-safe container and freeze for 2 hours. Whisk well and freeze once again until firm.

Banana Marijuana Ice Cream

Preparation time: 1minutes

Cooking time: 5 – 7 minutes

Makes: 20 – 25 servings

Ingredients:

- ½ stick butter

- 10 tablespoons sugar

- 1/8 teaspoon salt

- 6 tablespoons rum

- 0.7 ounce finely ground marijuana

- 36 ounces cream

- 30 ounces bananas, peeled, mashed

- 10 tablespoons honey

Directions:

1. Pour cream into a saucepan and place it over medium flame. When the cream is heated and simmering (it should be hot but not boiling), stir in the marijuana. Mix well and turn off the heat.

2. Add butter, sugar, and salt into another saucepan. Place over low heat to melt butter. Once butter melts, turn off the heat and stir until well combined.

3. Add the cream and bananas into the saucepan of butter mixture and whisk well.

4. Add honey and rum and beat until well combined.

5. Spoon the mixture into a freezer-safe container. Keep the container covered and place in the freezer. After 3 hours, remove the ice cream from the freezer and transfer the ice cream into a chilled bowl.

6. Whisk well. Cover with cling wrap and freeze until firm. 30 minutes before serving, remove the ice cream from the freezer and place in the refrigerator.

7. Serve.

Hash Fudge

Preparation time: 5 minutes

Cooking time: 10 minutes

Makes: 30 – 40 servings

Ingredients:

- 1 ½ cups heavy cream

- 2 teaspoons cornstarch

- 0.21 ounce hash

- 2 tablespoons vanilla extract

- 4 ounces chocolate, unsweetened, chopped

- 6 tablespoons butter

- 4 cups sugar

Directions:

1. Add milk, sugar, chocolate, and cornstarch and into a saucepan and stir. Place the saucepan over medium flame.

2. When the temperature of the mixture reaches 240° F, turn off the heat. Insert a cooking thermometer (also called candy thermometer) to check the temperature. Let the thermometer remain in it.

3. Add butter to a microwave-safe bowl. Microwave on high for a few seconds until the butter melts. Stir in the hash.

4. Continue cooking for another 30 seconds.

5. Pour this mixture into the bowl of chocolate mixture. Do not mix it.

6. When the temperature reaches 110° (make sure to work at this temperature because hurrying will only spoil your fudge), stir constantly until it gets difficult to move the spatula. It should take around 7-10 minutes. You need to be vigorous with the stirring.

7. Transfer into a baking dish. Let it cool completely. Cut into squares of the desired size. Store in an airtight container in the refrigerator.

8. Hash can be replaced with canna-butter if hash is unavailable.

No-Bake Cannabis Pumpkin Pie

Preparation time: 15 minutes

Cooking time: 0 minutes

Makes: 18 – 36 servings

Ingredients:

For crust:

- 2 cups pitted, chopped dates

- 1 cup shredded coconut

- ½ tablespoon nutmeg or cinnamon

- 4 tablespoons coconut oil

- 5 cups chopped nuts of your choice

- 2 tablespoons pumpkin pie spice

- 1/8 teaspoon salt

For filling:

- 3 cans pumpkin puree

- 3 cups maple syrup

- 4 cups cashews

- 6 tablespoons canna-coconut oil

Directions:

1. To make crust: Add dates, coconut nutmeg, coconut oil, nuts, pumpkin pie spice, and salt into the food processor bowl and process until well combined and sticky, with a few chunks.

2. Take 2 springform pans of about 10 to 12 inches each. Divide the crust mixture among the pans and press it onto the bottom and a little up the sides of the pan.

3. Place the pans in the freezer for 3 hours.

4. To make filling: Make the filling 10 minutes before pouring over the crust.

5. Add pumpkin puree, maple syrup, cashew, and canna-coconut oil into the food processor bowl and process until smooth.

6. Divide equally the filling among the piecrusts. Place the pans in the freezer for about 2 to 3 hours.

7. Remove from the freezer and place on your countertop for a few minutes before serving.

8. Cut into slices and serve. You can get 18 to 36 servings depending on the size of the slices.

Apple Pie

Preparation time: 20 minutes

Cooking time: 40 minutes

Makes: 8 – 10 servings

Ingredients:

For the crust:

- 1 2/3 cups all-purpose flour

- 1 tablespoon sugar

- 2 tablespoons chilled vegetable shortening, cut into cubes

- ½ teaspoon salt

- ½ cup canna-butter, chilled, cut into cubes

- 2 – 3 tablespoons chilled water

For the filling:

- 1 ½ pounds apples, cored, peeled, sliced

- 1 ½ tablespoons all-purpose flour

- A pinch salt

- A pinch ground nutmeg

- ½ tablespoon water

- 1 2/3 cups brown sugar

- 1 ½ teaspoons ground cinnamon

- ½ teaspoon granulated sugar

- 1 small egg, lightly beaten

- ½ tablespoon fresh lemon juice

Directions:

1. To make crust: Combine sugar, flour, and salt in a mixing bowl.

2. Add canna-butter and cut it into it the mixture until small crumbs are formed.

3. Pour chilled water, a tablespoon at a time, and mix well each time. Keep adding the water until you get smooth dough. Divide the dough into 2 portions, one portion slightly smaller than the other.

4. Wrap the dough balls in cling wrap and place in the refrigerator for 30 – 40 minutes.

5. Roll both the portions on the cling wrap itself and place it in the refrigerator until the filling is prepared.

6. Combine apples, brown sugar, and lemon juice in a bowl.

7. Combine flour, salt, cinnamon, and nutmeg in another bowl. Sprinkle this mixture over the apples and stir well. Let the juices release for 10 minutes.

8. Take a large pie pan and invert the bigger rolled dough over it. Peel off the cling wrap. Press the dough on the sides as well as the bottom of the pie pan.

9. Spread the apple mixture over the crust.

10. Remove the cling wrap from the smaller rolled dough and place the dough over the filling.

11. Seal edges of both the top and bottom dough together. Press the edges with a fork if you want to make a design. Make a few small slits on the top dough.

12. Add egg and water in a small bowl and whisk well. This is egg wash. Brush the egg wash over the top crust. Scatter sugar op top and place the pie pan in the refrigerator for 20 minutes.

13. Place the baking sheet in an oven that has been preheated to 425° F and bake for about minutes or until light brown.

14. Lower the temperature to 325° F and bake for 20 minutes or until golden brown on top.

15. Let the pie cool on your countertop for at least 10 – 15 minutes.

16. Cut into wedges and serve.

Cannabis-Infused Chocolate Cake

Preparation time: 10 minutes

Cooking time: 40 minutes

Makes: 12 servings

Ingredients:

- 1 ¼ cups all-purpose flour

- ½ cup canna-butter

- ½ cup boiling water

- 1 egg

- ¼ teaspoon salt

- 1 cup white sugar

- ½ cup buttermilk

- ¼ cup unsweetened cocoa powder + extra to sprinkle

- 1 teaspoon baking soda

Directions:

1. To prepare the oven and baking dish: Spray a baking dish with cooking spray. Sprinkle a little cocoa powder on the bottom of the dish. The oven has to be preheated to 350° F for about 10 minutes.

2. To mix dry ingredients: Add flour, cocoa, salt, and baking soda into a bowl and stir.

3. Add canna-butter and sugar into a mixing bowl. Beat with an electric hand mixer until creamy with a little fluff.

4. Add egg and beat well.

5. Add buttermilk and beat until just combined. Next, pour boiling water into the batter and beat for a minute.

6. Pour the batter into the prepared baking dish. Place the baking dish in the oven and bake for about 25 – 30 minutes. If a toothpick, when inserted in the middle of the cake, has no particles stuck on it when pulled out, your cake is ready. Otherwise, bake it for a few more minutes.

CHAPTER 7

SALAD & SOUP RECIPES

Salads

Cannabis Chicken Salad

Preparation time: 15 minutes

Cooking time: 0 minutes

Makes: 6 – 7 servings

Ingredients:

- 6 large chicken breasts, cooked, cut into bite-size pieces

- ½ cup diced red bell pepper

- ½ cup diced celery

- ½ cup diced onion

- Pepper to taste

- 2 tablespoons chopped fresh rosemary

- 2/3 cup canna- mayonnaise

- Salt to taste

Directions:

1. Add chicken, bell pepper, celery, onion, pepper, rosemary, canna-mayonnaise, and salt into a bowl and stir until well combined.

2. Cover and chill until use. This salad tastes great when chilled.

Thai Mango Salad

Preparation time: 15 minutes

Cooking time: 0 minutes

Makes: 2 servings

Ingredients:

- ½ tablespoon grated, fresh ginger

- Zest of a lime, grated

- 1 tablespoon canna-olive oil

- ½ tablespoon honey

- ½ jalapeño, deseeded, minced

- 1 small cucumber, peeled, diced

- 1 small red onion, thinly sliced

- 1 ripe mango, peeled, chopped

- ¼ bell pepper or any color, thinly sliced

- A handful fresh cilantro, chopped

- 1 tablespoon lime juice

- ½ tablespoon soy sauce

- 1 clove garlic, peeled, minced

Directions:

1. To make dressing: Add ginger, lime juice, soy sauce, garlic, cilantro, lime zest, honey, and canna-oil into a bowl and whisk well.

2. To make salad: Add jalapeño, cucumber, onion, mango, and bell pepper into another bowl and toss well.

3. Pour dressing over the salad and toss well.

Green Leafy Kale Salad with Brown Canna-Butter Vinaigrette

Preparation time: 15 minutes

Cooking time: 10 minutes

Makes: 2 – 3 servings

Ingredients:

For canna-butter vinaigrette:

- 3 tablespoons unsalted butter

- 3 tablespoons sherry vinegar or red wine vinegar

- 3 tablespoons canna-butter

- Salt to taste

For the salad:

- ¼ cup slivered almonds

- ½ pound Tuscan kale, discard hard ribs and stems, cut into bite-size pieces

Directions:

1. Place a pan over medium flame. Add almonds and toast them until you get a nice aroma. Turn off the heat and let it cool.

2. For canna-butter vinaigrette: Place a small saucepan over high flame. Add unsalted butter. Slowly the butter will melt, and in a few minutes, it will have bits of pieces in it and cook until the bites turn brown.

3. Stir in canna-butter. When it melts, turn off the heat.

4. Stir in the vinegar and salt.

5. Place kale in a bowl. Sprinkle almonds on top. Pour vinaigrette over it. Toss well and serve.

Caesar Salad

Preparation time: 10 minutes

Cooking time: 5 – 6 minutes

Makes: 6 – 8 servings

Ingredients:

- 2 heads romaine lettuce, rinsed, pat dried, chopped

- 8 cups bread cubes

- 6 cloves garlic, quartered lengthwise

- Pepper to taste

- ½ cup olive oil

- Salt to taste

For the dressing:

- 1 ½ cups mayonnaise

- ¾ cup grated parmesan cheese

- 2 teaspoons Worcestershire sauce

- 2 teaspoons Dijon mustard

- 2 tablespoons lemon juice

- 3 tablespoons cannabis-infused olive oil

- Pepper to taste

- 6 garlic cloves, minced

- Salt to taste

Directions:

1. To make dressing: Add mayonnaise, half the parmesan cheese Worcestershire sauce, Dijon mustard, lemon juice, canna-oil, pepper, garlic, and salt into a bowl and whisk well. Chill in the refrigerator until use.

2. To make bread croutons, pan-style: Place a large skillet over medium flame. Add olive oil. When the oil is heated, add garlic and stir-fry until brown. Remove garlic with a slotted spoon and place on a plate.

3. Add bread cubes into the pan. Stir-fry until light brown. Remove bread cubes from the pan and place in a bowl. Sprinkle salt and pepper over it. You can also make croutons by baking, given in the next recipe.

4. Place lettuce leaves, bread cubes, and remaining Parmesan cheese in a large bowl. Drizzle dressing over it. Toss well.

5. Serve.

Weed Salad

Preparation time: 10 minutes

Cooking time: 10 – 12 minutes

Makes: 4 servings

Ingredients:

- 4 thick slices bread, cubed

- Salt to taste

- 2 tablespoons lemon juice

- ¾ cup cannabis salad dressing

- ¼ cup grated parmesan cheese

- 6 teaspoons extra-virgin olive oil

- Pepper to taste

- 5 cups mixed lettuce leaves

- 2 cloves garlic, minced

- 2 cups grilled, chopped chicken

Directions:

1. To make croutons, baked style: Place croutons in a bowl. Drizzle oil over it. Sprinkle salt and pepper over it and toss well. Spread it over a baking sheet.

2. Place the baking sheet in an oven that has been preheated to 400° F and bake for 10 minutes or until golden brown.

3. Let the croutons cool on the baking sheet.

4. Transfer into an airtight container and store at room temperature until use.

5. Add salt, garlic, and lemon juice into a large bowl and whisk well.

6. Whisk in the cannabis dressing (few salad dressings are given in Chapter 3, use whatever suits you).

7. Add pepper to taste. Cover and set aside for a while for the flavors to blend in.

8. To make salad: Add lettuce, Parmesan, and chicken into the bowl of dressing and toss well.

9. Divide into plates. Scatter croutons on top and serve.

Strawberry Burrata Salad

Preparation time: 15 minutes

Cooking time: 0 minutes

Makes: 2 servings

Ingredients:

For vinaigrette:

- 1 ½ tablespoons strawberry puree

- ¾ tablespoon raw honey

- 6 tablespoons avocado oil

- 1 ½ tablespoons white balsamic vinegar

- 1/8 teaspoon pepper

- 10 drops CBD oil

For the salad:

- 3 cups packed arugula

- 2 tablespoons slivered, toasted almonds

- ½ cup sliced strawberries

- 2 balls (4 ounces each) burrata cheese

Directions:

1. To make dressing: Add strawberry puree, honey, vinegar, and pepper into a tall plastic glass that comes along with the immersion blender.

2. Using an immersion, blend the mixture until well combined; with the blender on, pour avocado oil in a thin drizzle. Mix until well combined.

3. Add CBD oil and mix well.

4. To make salad: Add arugula, strawberries, and burrata cheese into a bowl and toss well.

5. Add the dressing and toss well.

6. Scatter almonds on top and serve.

Spinach and Orange Salad with Grilled Salmon and Orange Vinaigrette

Preparation time: 20 minutes

Cooking time: 10 minutes

Makes: 8 servings

Ingredients:

For dressing:

- 2/3 cup orange juice

- 2 teaspoons olive oil

- 2 tablespoons soy sauce

- 3 teaspoons grated fresh ginger

- 1 teaspoon black pepper

- 2 tablespoons canna-olive oil

- 2 teaspoons toasted sesame oil

- 4 teaspoons agave nectar

- 1 teaspoon minced garlic

- Salt to taste

For salmon:

- 8 salmon fillets, 1 inch thick (6 ounces each)

- 4 teaspoons pepper

- 4 tablespoons fresh lemon juice

For salad:

- 16 ounces baby spinach

- 2 medium avocadoes, peeled, pitted cut into small cubes

- 1 red onion, thinly sliced

- 8 oranges, peeled, deseeded, cut into segments

- 2 small cucumbers, peeled, diced

Directions:

1. To make dressing: Add orange juice, olive oil, soy sauce, ginger, pepper, canna-oil, sesame oil, agave nectar, and garlic into a small jar. Shake the jar vigorously until well combined.

2. Set up your grill and preheat it to medium heat. Spray the rack with cooking spray.

3. Season salmon with pepper. Drizzle lemon juice over the fillets and place on the grill, with the skin side on top. Cook for 5 minutes. Flip sides and cook the other side for 5 minutes or until the fish flakes easily when pierced with a fork.

4. Peel off the skin from the fillets.

5. To make salad: Add spinach, avocado, onion, orange, and cucumber into a large bowl and toss well.

6. Pour dressing over it and toss well.

7. To serve: Place salad on individual serving plates. Place a salmon fillet on each plate, over the salad, and serve.

Cranberry Walnut Salad

Preparation time: 15 minutes

Cooking time: 0 minutes

Makes: 8 servings

Ingredients:

For apple cider cannabis vinaigrette:

- ½ cup canna-extra-virgin olive oil

- 1 teaspoon Dijon mustard

- ½ teaspoon garlic powder

- ½ teaspoon dried basil

- ½ teaspoon dried oregano

- Salt to taste

- 8 tablespoons apple cider vinegar

- 2 teaspoons light brown sugar

- ½ teaspoon freshly ground pepper or to taste

For salad:

- ¼ cup crushed walnuts

- 4 Fuji apples, cored, thinly sliced

- 2/3 cup dried cranberries

- 8 cups baby spinach

- 2/3 cup crumbled feta cheese

Directions:

1. To make dressing: Add mustard, garlic powder, herbs, salt, pepper, vinegar, and brown sugar into a blender. Blend until smooth.

2. With the blender machine running on low speed, pour canna-oil in a thin stream through the feeder tube. Blend until the mixture is emulsified.

3. Transfer the dressing into a bowl. Cover and set aside for a while for the flavors to meld.

4. To make salad: Add apples, walnuts, cranberries, spinach, and cheese into a bowl and toss well.

5. Pour dressing on top. Toss well and serve.

Canna-Quinoa Salad

Preparation time: 10 minutes

Cooking time: 20 minutes

Makes: 8 servings

Ingredients:

For dressing:

- 2 tablespoons olive oil

- 2 tablespoons canna-olive oil

- 2 teaspoons ground cumin

- 1 teaspoon ground black pepper

- 4 tablespoons fresh lemon juice

- 2 teaspoons minced garlic

- 1 teaspoon salt

For salad:

- 1 cup dry quinoa

- 4 tablespoons olive oil

- 2 cups corn, fresh or frozen

- 2 cups peas, fresh or frozen

- 2 cups frozen edamame

- 2 cans (15 ounces each) chickpeas

- ½ cup chopped red onion

- 2 cups chopped mint leaves

Directions:

1. To make salad: Cook quinoa following the directions on the package. You should have 4 cups cooked quinoa. You can use leftover cooked quinoa as well.

2. Place a large skillet over medium flame. Add olive oil. When the oil is heated, add corn, peas, edamame, and chickpeas and stir-fry for 5 to 6 minutes.

3. Turn off the heat and let it cool completely. Transfer the vegetables into a large bowl.

4. Add quinoa, onion, and mint and toss well.

5. While the quinoa is cooking, make the salad dressing. For this, add olive oil, lemon juice, garlic, salt, canna-oil, cumin, and pepper into a bowl and whisk well. Cover and set aside until the quinoa cools.

6. Pour dressing over the salad and toss until dressing is well combined with the salad.

7. Serve.

West Coast Garden Salad with Cannabis Olive Oil & Dill Dressing

Preparation time: 15 minutes

Cooking time: 0 minutes

Makes: 3

Ingredients:

For dressing:

- 6 tablespoons olive oil

- 3 tablespoons balsamic reduction

- 1 tablespoon canna-olive oil

- Pepper to taste

- 1 tablespoon chopped fresh dill

- Salt to taste

For salad:

- ½ head red lettuce, chopped

- ½ large beet, peeled, cut into cubes

- ¼ cup chopped fresh dill

- ¼ cup salted pumpkin seeds

- 1 medium carrot, peeled, grated

- ½ cup grape tomatoes

- ¼ cup crumbled feta cheese

Directions:

1. To make dressing: Add olive oil, balsamic reduction, canna-olive oil, pepper, dill, and salt into a bowl and whisk well. Cover and set aside for a while for the flavors to mingle.

2. To make salad: Add lettuce, beet, dill, pumpkin seeds, carrot, tomatoes, and feta cheese into a bowl and toss well.

3. Pour dressing on top. Toss well and serve.

Almond Apricot Chicken Salad

Preparation time: 20 minutes

Cooking time: 10 minutes

Makes: 8 servings

Ingredients:

For dressing:

- 2 cups sour cream
- ½ cup canna-olive oil
- 1 cup mayonnaise
- 4 teaspoons grated lemon zest
- 2 teaspoons salt or to taste
- 1 teaspoon pepper or to taste
- 2 tablespoons lemon juice
- 4 teaspoons Dijon mustard
- 1 ½ teaspoons dried savory

For salad:

- 1 package (16 ounces) spiral pasta
- 6 cups coarsely chopped broccoli florets
- 1 cup chopped green onions
- 12 ounces dried apricots
- 5 cups cooked, diced chicken
- 1 cup chopped celery
- 1 ½ cups sliced almonds

Directions:

1. To make dressing: Add sour cream, canna-oil, mayonnaise, lemon zest, salt, pepper, lemon juice, mustard, and dried savory into a bowl and whisk well. Cover and set aside until the pasta is cooked.

2. To make salad: Cook the pasta following the directions on the package. Add past 4 minutes before draining the pasta. Drain the pasta in a colander and place it under cold running water for a few minutes until it cools. Let it remain in the colander for 10 minutes.

3. Add pasta, broccoli, green onions, chicken, and celery into a bowl and toss well.

4. Combine the salad and dressing and refrigerate until use. Keep it covered in the refrigerator.

5. Meanwhile, place a pan over medium flame. Add almonds and toast the almonds to the desired doneness.

6. Divide salad into plate. Scatter almonds on top and serve.

CBD Salad

Preparation time: 20 minutes

Cooking time: 0 minutes

Makes: 4 – 8 servings

Ingredients:

For salad:

- 6 – 8 cups mixed greens or arugula

- 4 thick pieces bacon or smoked seitan bacon, cooked

- 1 cup crumbled goat's cheese

- 2 cups chopped walnuts

- 1 cup fresh or dried cranberries

- 2 apples cored, cut into pieces

For dressing:

- 4 tablespoons CBD oil

- 1 cup olive oil

- 4 tablespoons apple cider vinegar

- 4 teaspoons maple syrup

- 3 tablespoons poppy seeds

- 1 teaspoon salt

- 1 teaspoon ground mustard

- 2 – 3 tablespoons poppy seeds

- ½ teaspoon freshly ground pepper

- 2 tablespoons finely chopped shallot

- 1 tablespoon hemp seeds or hemp hearts

Directions:

1. To make dressing: Add CBD oil, shallots, and vinegar into a bowl and whisk well.

2. Whisk in the maple syrup, mustard, salt, pepper, poppy seeds, and hemp hearts.

3. Cover and set aside for at least 15 minutes.

4. Meanwhile, make the salad by tossing together apple, bacon, greens, and walnuts.

5. Pour dressing over the salad. Toss well and serve.

Multiple Sclerosis (MS) Arugula Goat-Cheese Salad with Cannabis Citrus Vinaigrette

Preparation time: 15 minutes

Cooking time: 0 minutes

Makes: 2 servings

Ingredients:

For salad:

- ½ cup baby spinach

- A handful cauliflower florets, sliced

- 3 tablespoons shredded carrots

- 1/8 cup crumbled goat cheese

- 1 ½ cups arugula

- ½ cup baby kale leaves

- ¼ cup sliced shiitake mushrooms

- 1 tablespoon chopped pine nuts

- ½ clementine, separated into segments

For citrus vinaigrette:

- 1 tablespoon flaxseed oil

- 2 tablespoons canna-extra-virgin olive oil

- Pepper to taste

- Juice of ½ clementine

- 1 tablespoons apple cider vinegar

- ½ tablespoon clover honey

- 1 tablespoon fresh lemon juice

- Salt to taste

Directions:

1. Dry the greens by patting with paper towels or a kitchen towel.

2. Add all the greens, mushrooms, carrot, cauliflower, clementine, pine nuts, and goat cheese into a large bowl and toss well.

3. To make dressing: Combine clementine juice, lemon juice, apple cider vinegar, flaxseed oil, and seasonings into a blender and blend until well combined.

4. With the blender machine running, pour canna-oil in a thin stream and blend until emulsified. If you want thick dressing, pour some more oil.

5. Pour dressing over the salad. Toss well and serve.

Infused Fruit Salad

Preparation time: 15 minutes

Cooking time: 0 minutes

Makes: 4 servings

Ingredients:

- 1 cup sliced strawberries

- ½ cup fresh cherries pitted

- ½ cup blueberries

- 2 oranges, peeled, separated into segments, deseeded

For dressing:

- 4 teaspoons canna-olive oil

- 2 teaspoons basil-infused sesame oil

Directions:

1. Toss together the berries, cherries, and oranges in a bowl.

2. Pour basil-infused sesame oil canna-olive oil over the fruits and toss well.

3. Chill for a few hours. It can be served as it is or as a topping for pancake, waffles, or just over anything.

THC Tuna Salad

Preparation time: 10 minutes

Cooking time: 20 minutes

Makes: 2 servings

Ingredients:

- 2 – 3 droppers THC tincture

- 1/8 teaspoon garlic powder

- ½ tablespoon dried parsley

- A pinch dried, minced onion flakes

- ½ tablespoon grated parmesan cheese

- 1/8 teaspoon curry powder

- 1 ½ tablespoons sweet pickle relish

- 3 tablespoon mayonnaise

- ½ can (from a 7 ounces can) white tuna, drained, flaked

- ½ teaspoon dried dill weed

Directions:

1. Combine THC tincture, garlic powder, parsley, onion flakes, Parmesan cheese, curry powder, sweet pickle relish, and dill in a bowl.

2. Add tuna and mix well.

3. Serve as a filling for sandwiches or over crackers or over lettuce leaves.

Soups

Marijuana Tortilla Soup with Vegetables

Preparation time: 20 minutes

Cooking time: 20 minutes

Makes: 6 servings

Ingredients:

- 2 medium corn tortillas, cut into thin strips, as thin as matchsticks

- 1 ½ - 2 cups cooked chicken or turkey

- ½ cup water

- ½ cup shredded Monterey Jack cheese

- 2 cups chicken or turkey stock

- ½ can (from a 15 ounces can) corn with its liquid

- 1 cup cannabis salsa

- Any other vegetables of your choice (optional)

To serve:

- A handful fresh cilantro, chopped

- Lime slices

Directions:

1. Place tortilla strips on a baking sheet without overlapping.

2. Bake the tortilla strips in a preheated oven at for about 15 minutes or until light brown in color. Turn off the oven and let the tortilla strips cool on your countertop.

3. Once cooled, set aside one-third of the tortilla strips and add the remaining strips into a blender and blend until crushed. Do not grind it to fine powder.

4. Add salsa, chicken, water, stock, and crushed tortillas into a soup pot. Place the pot over medium-high flame. When it comes to a boil, reduce the heat and cook on low.

5. Stir in chicken and corn and any vegetables if using. Cook until vegetables are tender.

6. Add cheese and stir until cheese melts.

7. Ladle into soup bowls. Garnish with retained tortilla strips and cilantro. Serve with lime slice.

Marijuana French Onion Soup Au Gratin

Preparation time: 20 minutes

Cooking time: 60 – 70 minutes

Makes: 8 servings

Ingredients:

- 4 tablespoons unsalted butter

- 2 large sweet onions like Vidalia

- 4 large leeks, use only white and pale green parts

- 2 tablespoons cannabis olive oil

- 12 cups beef stock

- 2 bay leaves

- Salt to taste

- 2 2/3 cups shredded gruyere or Swiss cheese

- 2 tablespoons olive oil

- 4 large shallots

- 2 tablespoons minced garlic

- 1 cup dry sherry

- 2 teaspoons Worcestershire sauce

- 2 teaspoons balsamic vinegar

- 8 slices French baguette (one-day-old bread)

- Pepper to taste

Directions:

1. Place a soup pot over medium flame. Add butter and oil. When it heats, add onion, leeks (white part), and shallots and stir-fry for a few minutes until the onions start to look light brown in color.

2. Lower the flame and cook until onions are caramelized, stirring occasionally. Scrape the bottom of the pot to remove any particles that are stuck.

3. Stir in the garlic and cook for a couple of minutes.

4. Pour sherry into the pot and cook on medium-high heat. Scrape the bottom of the pot to remove any particles that may be stuck.

5. Cook until nearly dry. Stir in canna-oil. Once well combined, pour stock and Worcestershire sauce along with bay leaves. When the soup begins to boil, lower the heat and cook for about 45 minutes.

6. Discard the bay leaves. Add balsamic vinegar. Turn off the heat.

7. Take 8 ovenproof bowls and place them on a baking sheet.

8. Divide the soup among them. Divide equally half the cheese among the bowls, and scatter on top. Place a slice of baguette in each bowl. Sprinkle remaining cheese on top.

9. Set the oven to broil mode. Place the baking sheet in the oven and broil until cheese melts.

10. Serve right away.

Cannabis Chicken Noodle Soup

Cooking time: 30 minutes

Makes: 8 servings

Ingredients:

- 1 pound chopped, cooked chicken breasts

- 8 cups chicken broth

- 3 ½ cups vegetable broth

- 2 cups sliced carrots

- 1 cup chopped onions

- 2 celery stalks, diced

- ½ cup peas

- Salt to taste

- 12 droppers cannabis tincture

- Pepper to taste

- 2 tablespoons butter

- 1 teaspoon dried basil

- 1 teaspoon dried oregano

- 3 cups uncooked egg noodles

Directions:

1. Place a soup pot over medium flame.

2. Add butter. Once butter melts, add onions and celery and sauté until the onions are translucent.

3. Add chicken, carrots, oregano, basil, egg noodles, peas, cannabis tincture, and chicken stock.

4. Cook until the vegetables and noodles are tender.

5. Add salt and pepper to taste.

6. Ladle into soup bowls and serve hot.

Greek Lemon Chicken Soup

Preparation time: 10 minutes

Cooking time: 20 – 25 minutes

Makes: 8 servings

Ingredients:

- 10 cups chicken or turkey stock

- 4 cups cooked rice, divided

- 4 egg yolks

- Salt to taste

- 2 chicken breasts, skinless, boneless

- 2 tablespoons canna-olive oil

- Pepper to taste

- ½ cup lemon juice

Directions:

1. Pour stock into a soup pot and place the pot over high flame. When the stock begins to boil, drop the chicken in it and cook until chicken is well cooked.

2. Remove chicken with a slotted spoon. Let the stock simmer on low heat.

3. When chicken is cool enough to handle, shred with a pair of forks and keep it in a bowl.

4. Add 1 cup cooked rice and 2 cups of stock into a blender and blitz until smooth.

5. Let the blender be running, pour yolks, canna-olive oil, and lemon juice and blend until smooth.

6. Pour into the simmering soup. Stir often. Also, add in the chicken and 3 cups rice and simmer for about 12 – 13 minutes or until you are satisfied with the thickness.

7. Ladle into soup bowls and serve.

Vegan Split Pea Soup

Preparation time: 1 hour and 15 minutes

Cooking time: 60 – 90 minutes

Makes: 3 – 4 servings

Ingredients:

- 2 ½ cups vegetable broth

- 2 cups dried split peas, rinsed, soaked in water for an hour

- ½ tablespoon canna- extra- virgin olive oil

- 1 ½ tablespoon canna-coconut oil

- 4 cloves garlic, finely minced

- 1 bay leaf

- Pepper to taste

- 2 – 3 cups water

- 1 large carrot, cut into bite-size chunks

- ½ large white onion, finely diced

- 1 ½ -2 tablespoons white miso paste or to taste

- ½ teaspoon dried thyme

- Salt to taste

Directions:

1. Place a soup pot over medium flame. Add canna- extra-virgin olive oil. When the oil is heated, add onion and sauté until onions are translucent.

2. Add garlic and sauté until aromatic. Add carrot and split peas. Add salt and pepper to taste.

3. Stir in miso, thyme, canna- coconut oil, and bay leaf and mix well.

4. Stir in broth and water. Let it come to a boil.

5. Lower the heat and keep the pot cover with a lid. Simmer until split peas are cooked. Stir occasionally. Discard the bay leaf

6. If you have an instant pot, you can cook it in it. It is much quicker.

7. Ladle into soup bowls and serve.

Tomato Soup with Carrots and Celery

Preparation time: 15 minutes

Cooking time: 30 minutes

Makes: 8 servings

Ingredients:

- ½ cup canna-butter

- 2 cans (8 ounces each) tomato sauce

- ¼ cup fresh basil

- Salt to taste

- 6 tablespoons butter

- Pepper to taste

- 2 ½ cups chicken broth

- 2 tablespoons chopped oregano

- 3 cups heavy whipping cream

- 30 baby carrots, thinly sliced

- 6 cloves garlic, chopped

Directions:

1. Place a soup pot over a medium-low flame. Add butter and let it melt.

2. Add all the vegetables and stir-fry until tender.

3. Add herbs, broth, and tomato sauce. Raise the heat to medium and cook for about 20 minutes. Turn off the heat. Cool for a while.

4. Blend the soup in batches and pour it back into the pot. Add cream and place the pot over medium flame. Heat thoroughly.

5. Stir in canna-butter. Once butter melts, turn off the heat.

6.   Ladle into soup bowls and serve.

Hearty Vegan Winter Vegetable Soup

Preparation time: 20 minutes

Cooking time: 20 minutes

Makes: 5 servings

Ingredients:

- ½ tablespoon olive oil

- ½ medium onion, finely diced

- ½ large carrot, finely diced

- 6 tablespoons red wine

- ½ can (from a 16.5 ounces can) diced tomatoes with its juice

- 1 ¼ cups water

- ¼ small head green cabbage, thinly sliced

- 1 ½ tablespoons soy sauce

- ½ teaspoon dried oregano

- Hot sauce to taste

- ½ large leek, use only white part, thinly sliced

- ½ stalk celery, finely diced

- ½ tablespoon minced garlic

- ½ tablespoon balsamic vinegar or cider vinegar

- 2 cups vegetable stock

- 1 cup chopped, mixed vegetables

- ½ teaspoon pepper

- 0.03 ounce decarbed hash or kief, finely ground

- Salt to taste

Directions:

1. Place a soup pot over medium-high flame. Add oil and let it heat. Add onion and leeks and cook until light brown. Stir in carrots and celery.

2. Cook for about 5 minutes. Stir in garlic and cook for a few seconds until you get a nice aroma.

3. Add wine and vinegar and scrape the bottom of the pot to remove any particles that may be stuck.

4. Stir in the tomatoes, cabbage, and mixed vegetables. Pour stock and water and cook on low until veggies are soft.

5. Add soy sauce, oregano, pepper, salt, and kief and cook for a couple of minutes.

6. Ladle into soup bowls and serve.

Cream of Cannabis Soup

Preparation time: 10 minutes

Cooking time: 20 minutes

Makes: 8 servings

Ingredients:

- 6 cups vegetable stock
- ½ cup chopped red onion
- 2 tablespoons flour
- 4 cloves garlic, finely chopped
- Pepper to taste
- 2 cups chopped broccoli
- 2 cups sliced celery
- 4 cups heavy cream
- ½ cup chopped cilantro
- Canna-butter, to suit your requirements

Directions:

1. Place a saucepan over medium flame. Add canna-butter and let it melt.
2. Add onion, garlic, and celery and cook for a couple of minutes.
3. Add flour and stir constantly for about a minute.
4. Stirring constantly, pour stock into the pot. Add broccoli and celery and cook until tender.
5. Add heavy cream and heat thoroughly.
6. Add pepper and stir.

7. Ladle into soup bowls. Garnish with cilantro and pepper and serve.

Cannabis Tomato Soup

Preparation time: 15 minutes

Cooking time: 5 minutes

Makes: 8 servings

Ingredients:

- 6.6 pounds plum tomatoes, chopped into chunks
- 6 cloves garlic, finely chopped
- 1 teaspoon pepper
- 6 tablespoons balsamic vinegar
- 4 tablespoons Italian spice blend
- 4 handfuls fresh cannabis leaves
- 2 tablespoons canna-olive oil
- Juice of a lemon
- Salt to taste
- 1 red onion, chopped
- 2 handfuls fresh cilantro, chopped
- 2 handfuls fresh dill, chopped
- ¼ cup crumbled feta cheese

Directions:

1. Add tomatoes, salt, lemon juice, Italian spice blend, garlic, and pepper into a bowl and stir well.

2. Cover and set aside for an hour.

3. Add into a blender along with vinegar, garlic, onion, and canna-oil and blend until nearly smooth. Pour into 8 bowls.

4. Place cannabis leaves in a pan. Cover with water (about ¼ inch in height from the bottom of the pan). Let the leaves cook for about 5 minutes.

5. Divide the cannabis leaves among the bowls. Sprinkle dill, cilantro, and feta cheese on top, in each bowl, and serve.

Green Cannabis Soup

Preparation time: 15 minutes

Cooking time: 1 hour and 15 – 20 minutes

Makes: 8 – 10 servings

Ingredients:

- 4 carrots, cubed

- 6 cloves garlic, minced

- 1 purple cabbage, sliced

- 2 stalks celery, sliced

- 2 jalapeño peppers, sliced

- Salt to taste

- 4 sweet potatoes, peeled, cubed

- 1 cup chopped parsley

- 2 red onions, chopped

- 4 leeks, sliced

- 2 cups diced plum tomatoes

- 4 bell peppers, cut into squares

- 6 tablespoons olive oil

- Pepper to taste

- 4 tablespoons canna-butter

- Water, as required

Directions:

1. Place a soup pot over medium flame. Add olive oil and canna-butter and let the butter melt.

2. Stir in onion and garlic. Add cabbage, celery, sweet potatoes, celery, jalapeños, leeks, tomatoes, and bell peppers after the onion turns translucent.

3. Add 6 – 8 cups water or add more if desired. Let it come to a boil.

4. Reduce the heat and simmer for 60 – 80 minutes. Turn off the heat.

5. Ladle into soup bowls. Sprinkle parsley on top and serve.

Weed Ramen Noodles

Preparation time: 5 minutes

Cooking time: 30 minutes

Makes: 2 servings

Ingredients:

- ½ stick butter

- 2 servings Ramen kimchi noodles

- 4 cups water

- 0.1 – 0.14 ounce weed, ground

Extra flavorings: Optional

- Hot sauce

- Cheese

- Oregano

- Chili etc.

Directions:

1. Place a pot over medium flame. Add water and heat until slightly hot but do not boil. Stir in the weed.

2. Add butter and simmer on medium-low flame until the weed turns slightly brown in color. Do not boil. If it starts boiling, turn off the heat for a while.

3. Add the ramen noodles with the seasoning that comes with it and the optional flavorings.

4. Simmer until the noodles are cooked.

5. Ladle into soup bowls and serve.

Vegan Creamy Cannabis-Infused Potato Soup

Preparation time: 20 minutes

Cooking time: 30 minutes

Makes: 6 – 8 servings

Ingredients:

- 1 ½ tablespoons canna-oil

- 2 cloves garlic, minced

- 1 ½ pounds potatoes, peeled, cubed

- 1 ½ cups vegetable broth

- Salt to taste

- ½ large onion, chopped

- 1 ½ medium carrots, peeled, sliced

- ¼ teaspoon dried thyme

- 1 cup full-fat coconut milk

Directions:

1. Place a soup pot over medium flame. Add canna-oil and let it heat.

2. Once the oil is heated, add onion and garlic and cook until soft. Stir often.

3. Stir in potatoes, thyme, carrots, and broth. Once potatoes are cooked, turn off the heat.

4. Blend with an immersion blender until smooth. You can also blend it in a food processor or blender. There should be no lumps of the vegetables, so you need to blend it well.

5. Add coconut milk and blend once again. Add salt and stir.

6. Ladle into soup bowls and serve.

Cannabis Mushroom Soup

Preparation time: 15 minutes

Cooking time: 20 minutes

Makes: 2 servings

Ingredients:

- A handful shiitake mushrooms
- 1 cup heavy cream
- 2 small cloves garlic, peeled, minced
- 1 teaspoon flour (optional)
- Pepper to taste
- Canna-butter, as required
- 1 cup vegetable broth
- 1 small onion, chopped
- ½ tablespoon olive oil
- Salt to taste

Directions:

1. Place a pan over medium flame. Add canna-butter, as per your requirement, and let it melt.

2. Add mushrooms and cook until brown. Transfer into a bowl, including the cooked liquid.

3. Let the pan dry. Add oil and let it heat over medium-high flame. Add onion and cook until translucent.

4. Stir in garlic and cook for a few seconds until you get a nice aroma.

5. Pour vegetable broth and cream. Let it come to a boil, stirring often.

6. If you are using flour, mix it with a tablespoon of water and pour into the pan. Stir constantly until thick. Add salt and pepper to taste.

7. Ladle into soup bowls and serve.

Cannabis-Infused Bone Broth

Preparation time: 10 minutes

Cooking time: 24 – 48 hours

Makes: 8 – 10 servings

Ingredients:

- 2 ½ pounds beef bones

- 1 onion, cut into thick slices

- 1 stalk celery, cut into 1 inch pieces

- 1 large bay leaf

- 2 tablespoons apple cider vinegar

- 1 carrot, cut into chunks

- 2 cloves garlic, peeled

- Salt to taste

- 0.07 ounce cannabis, chopped

- Pepper to taste

Directions:

1. To prepare the oven: The oven has to be preheated to 240° F for about 20 minutes.

2. Place bones, vegetables, and cannabis on a rimmed baking sheet and toss well. Spread it all over the baking sheet and place the baking sheet in the oven for 40 minutes.

3. Transfer the bones and vegetables along with cannabis into a crockpot. Add seasoning, bay leaf, and apple cider vinegar and stir. Pour enough water to cover the ingredients in the pot by 3 – 4 inches (above the ingredients).

4. Set the crockpot on "Low" and cook for 24 – 48 hours.

5. Discard the bones, cannabis, and vegetables. You can strain the broth into a jar

6. Ladle into soup bowls and serve. Store leftovers in the refrigerator. It can last for 7 – 8 days.

Butternut Squash Soup with CBD Drizzle

Preparation time: 10 minutes

Cooking time: 40 – 45 minutes

Makes: 3 servings

Ingredients:

- 1 ½ tablespoons olive oil

- ½ cup cashews

- 3 cups peeled, cubed butternut squash

- 1 tablespoon minced, fresh ginger

- ¾ teaspoon ground coriander

- ½ teaspoon curry powder

- Salt to taste

- ½ teaspoon canna- coconut oil

- 1 medium shallot, peeled, finely chopped

- 2 small cloves garlic, minced

- 2 cups vegetable stock

- ¾ teaspoon ground cumin

- ½ tablespoon maple syrup

- ½ teaspoon ground turmeric

- ¾ cup coconut milk, divided

- 3 eggs, poached

Directions:

1. Place a soup pot over medium flame. Add olive oil and let it heat. Add shallots and sauté for a couple of minutes.

2. Stir in cashews. Once the cashews turn light brown, stir in the garlic.

3. Stir-fry for a few seconds until you get a nice aroma. Stir in butternut squash, salt, and all the spices. Keep stirring until you get a nice aroma.

4. Stir in the stock and maple syrup. When the soup begins to boil, lower the heat and cook covered until squash is fork-tender.

5. Stir in ½ cup coconut milk. Turn off the heat. Let it cool for a few minutes. Blend the soup until smooth.

6. Pour the blended soup back into the pot. Heat thoroughly just before serving. Also, poach the eggs just before serving.

7. Add canna-coconut oil and ¼ cup coconut milk into a bowl and whisk well.

8. Ladle the soup into soup bowls. Trickle the coconut milk mixture on top. Swirl lightly. Top with a poached egg in each bowl and serve.

CHAPTER 8

SNACK RECIPES

Weed Deviled Eggs

Preparation time: 5 minutes

Cooking time: 15 minutes

Makes: 6 servings

Ingredients:

- 3 large eggs

- 2 – 3 tablespoons mayonnaise

- Salt to taste

- Pepper to taste

- Paprika to taste

- 2 tablespoons canna-oil

- ½ tablespoon minced green onion

Directions:

1. Place eggs in a saucepan. Cover with water and place the saucepan over high flame.

2. When it comes to a boil, reduce the flame and let it simmer for 10 to 11 minutes. Turn off the heat.

3. Drain off the water and pour cold water into the saucepan. Let the eggs cool completely.

4. Remove the shells and cut each into 2 halves, lengthwise. Carefully scoop out the yolks and place in a bowl. Keep the whites on a serving platter, with the cavity (yolk) side facing upwards.

5. Add canna-oil, mayonnaise, salt, green onion, and pepper and mash well.

6. Transfer this mixture into a piping bag. Pipe the mixture into the cavities in the whites. Sprinkle paprika on top. Chill until use.

Weez-Its

Preparation time: 5 minutes

Cooking time: 25 minutes

Makes: 2 – 3 servings

Ingredients:

- 2 ounces canna-oil

- 2 cups Cheez-Its

Directions:

1. Add Cheez-Its into a bowl. Drizzle canna-oil over it and toss gently.

2. Line a baking dish with aluminum foil. Add the Cheez-Its into the baking dish and spread it evenly.

3. Place the baking sheet in an oven that has been preheated to 250° F and bake for 10 minutes or until golden brown.

4. Cool completely and serve.

Jalapeno Pot Poppers

Preparation time: 30 minutes

Cooking time: 15 minutes

Makes: 24 servings

Ingredients:

- 2 small links chorizo (optional)

- 1 cup grated mozzarella cheese

- 24 medium jalapeño peppers

- 6 large eggs, beaten

- Salt to taste

- Vegetable oil, to fry, as required

- 1 cup grated Monterrey Jack cheese

- 2 teaspoons dried oregano

- 0.07 ounce decarbed kief or hash, finely crumbled

- 2 cups dried breadcrumbs

- Pepper to taste

Directions:

1. Place a skillet over medium flame. Add chorizo and cook until brown, breaking it as you stir and cook.

2. Add mozzarella and Jack cheese and mix well. Remove from heat and let it cool for a while.

3. Slice off a thin part from the top of the jalapeño peppers. Remove the seeds and membranes. You should be able to stuff the peppers with the filling.

4. Make 24 equal portions of the cheese mixture. Also, make 24 equal portions of the kief.

5. Place a portion of the cheese mixture on your palm. Flatten it slightly and place a portion of kief on it. Bring together the edges of the cheese mixture to enclose the kief. Make it elongated in shape and stuff this inside a jalapeño pepper.

6. Press together the top edges of the pepper so that the filling goes right inside.

7. Repeat steps 5 – 6 and fill the remaining peppers.

8. Place breadcrumbs, pepper, and salt in a bowl and stir.

9. Dip the peppers in egg, one at a time. Shaking off extra egg, dredge in breadcrumbs mixture.

10. Repeat the previous step once again with each pepper and place on a baking sheet.

11. Place a large skillet over medium flame. Pour enough oil such that it is about 2 inches in height from the bottom of the pan. When the oil is well heated but not smoking, 325° F, carefully drop a few stuffed and breaded peppers in the oil. Cook until golden brown.

12. Remove the peppers with a slotted spoon and set aside on a plate lined with paper towels.

13. Fry the remaining peppers in batches.

14. Serve with a dip of your choice.

Marijuana-Infused, 3 Layered Popsicle

Preparation time: 5 minutes

Cooking time: 0 minutes

Makes: 8 – 9 servings

Ingredients:

For blueberry layer:

- 1 cup frozen blueberries

- 1/8 cup full-fat coconut milk

- 1 tablespoon maple syrup or stevia to taste

- ½ teaspoon CBD oil

- 1 teaspoon canna-coconut oil

For coconut milk layer:

- Maple syrup to taste

- ½ can full-fat coconut milk

- For strawberry layer:

- 6 ounces strawberries, chopped

Directions:

1. Add blueberries, coconut milk, maple syrup, CBD oil, and canna-coconut oil into the food processor bowl or blender and process until well combined and smooth.

2. Pour 2 tablespoons of the blueberry mixture into 8 – 9 Popsicle molds. Freeze for about an hour or until firm.

3. Rinse the blender.

4. To make coconut milk layer: Add coconut milk and maple syrup into a blender and blend until smooth.

5. Pour into the Popsicle molds and place the molds in the freezer for 15 minutes.

6. Rinse the blender once again.

7. To make strawberry layer: Add strawberries into a blender and blend until smooth.

8. Pour over the coconut layer. Insert the Popsicle sticks in the molds. Freeze until firm.

Watermelon and Lime Cannabis-Infused Popsicles

Preparation time: 10 minutes

Cooking time: 0 minutes

Makes: 3 – 4 servings

Ingredients:

- 1 ½ cups cubed watermelon, seedless

- 2 teaspoons canna-honey

- 1 ½ tablespoons coconut oil

- ½ tablespoon lime juice

Directions:

1. Add watermelon, canna-honey, coconut oil, and lime juice into a blender and blend until pureed.

2. Pour into 3 – 4 Popsicle molds. Insert sticks in them. Freeze until firm.

Baked Apricot Brie

Preparation time: 15 minutes

Cooking time: 30 – 40 minutes

Makes: 24 – 28 servings

Ingredients:

- 2 Brie rounds (8 ounces each)

- 2 cups canna-butter, melted

- ¼ teaspoon salt

- 1 package (16 ounces) filo dough sheets, thawed

- 2/3 cup apricot preserve

- ½ cup roasted, salted, chopped almonds (optional)

Directions:

1. Cut off the outer rind of the Brie rounds if desired.

2. Add canna-butter into a saucepan. Place over low heat until it melts. Turn off the heat.

3. Unfold the filo dough on your countertop.

4. Pull out 2 filo sheets and place on your countertop, slightly overlapping. You should have a rectangle of approximately 24 x 17 inches.

5. Brush melted butter over the filo sheets.

6. Place 2 more filo sheets over this, similarly, overlapping slightly.

7. Repeat step 5-6. You should have a stack with at least 3 to 4 layers. The topmost layer should be brushed with butter.

8. Place one Brie round in the center of the filo rectangle.

9. Smear 5-6 tablespoons of the apricot preserve over the Brie. You can use orange marmalade instead of apricot preserve or any other preserve of your choice.

10. Sprinkle salt and half the almonds over the Brie.

11. Fold the edges of the filo layers over the Brie to cover it completely. Press together the edges to seal.

12. Do this with the other Brie as well (steps 2 to 11).

13. Place the Brie filled filo's on a lined baking sheet, with the seam side facing down.

14. Place the baking sheet in an oven that has been preheated to 375° F, for 30 to 40 minutes or until golden brown. You may find that Brie oozing out of the pressed edges.

15. Remove the baking sheet from the oven and let it cool for 15 minutes.

16. Cut into slices. It is best served over crackers.

Cannabis-Infused Salted Caramel Popcorn

Preparation time: 10 minutes

Cooking time: 15 minutes

Makes: 2 – 3 servings

Ingredients:

- 1 ½ quarts popped popcorn

- 2 tablespoons butter

- 2 tablespoons honey

- ¼ + 1/8 teaspoon baking soda

- 1 tablespoons canna-butter

- ½ cup dark brown sugar

- 1 teaspoon salt

- ½ teaspoon vanilla extract or maple extract

Directions:

1. To prepare oven and baking sheet: Place a sheet of parchment paper over a large baking sheet. Make sure that your oven is preheated to 225° F.

2. Place popcorn in a bowl

3. Add canna-butter and butter into a saucepan. Place the saucepan over medium flame.

4. Add honey, brown sugar, and ¼ teaspoon salt and keep stirring. Slowly the mixture will start turning light brown.

5. Turn off the heat. Add baking soda and vanilla extract and stir constantly. Quickly pour it over the popcorn and toss well immediately. You need to be very quick else; the caramel will become hard.

6. Transfer the popcorn onto the baking sheet. Season with ¾ teaspoon salt.

7. Place the baking sheet in the oven and bake for about 30 minutes, stirring halfway through baking.

8. Take out the baking sheet from the oven and let it cool for 5 – 8 minutes.

9. Serve. Leftovers can be stored in an airtight container. It can last for a couple of days.

Monster Munchie Balls

Preparation time: 5 minutes

Cooking time: 5 – 6 minutes

Makes: 6 – 8 servings

Ingredients:

- ¾ cup canna-butter

- 2 tablespoons chunky peanut butter

- 1 tablespoon cocoa powder

- 1 ½ cups rolled oats

- 1 ½ tablespoons honey

Directions:

1. Add canna-butter into a saucepan and place the saucepan over low flame.

2. Stir in peanut butter, cocoa, oats, and honey. Stir constantly.

3. Transfer into a baking dish and spread it evenly. Freeze for 10 minutes.

4. Line an airtight container with parchment paper or wax paper.

5. Make small balls of the mixture (about 1 inch) place in the container. Close the lid and refrigerate until use.

Cinnamon Muffins

Preparation time: 15 minutes

Cooking time: 20 minutes

Makes: 6 servings

Ingredients:

- 1 ½ cups flour

- 1 egg

- 1 cup cane sugar

- 1 ½ tablespoons baking powder

- ¼ teaspoon ground nutmeg

- ½ tablespoon ground cinnamon

- 2 ½ tablespoons canna-butter

- ½ cup butter, melted

- ½ cup milk

- 2 ½ tablespoons shortening

- ½ teaspoon salt

Directions:

1. Grease a 6 counts muffin pan with some cooking spray. Place muffin liners in each cup.

2. Add canna-butter, ½ cup sugar, and shortening in a bowl. Beat until creamy.

3. Beat in the egg, salt, nutmeg, and baking powder.

4. Add a little flour at a time alternating with a little milk, and beat until just combined. Do this until all the milk and flour is added. You may have a few small lumps, but that is alright but do not over-beat.

5. Pour into the muffin cups, equal quantity in each cup.

6. Place the muffin pan in an oven that has been preheated to 350° F for 20 minutes or until a toothpick, when inserted in the middle of the muffins, comes out without any particles stuck on it.

7. Remove the muffin pan from the oven and let it cool for 15 minutes. Remove the muffins from the pan and place on a serving platter.

8. In the meantime, add ½ cup sugar and cinnamon into a bowl and stir.

9. Drizzle melted butter over the muffins. Scatter cinnamon sugar over the muffins and serve.

Weed Mozzarella Sticks

Preparation time: 20 minutes

Cooking time: 10 minutes

Makes: 2 – 3 servings

Ingredients:

- 1 cup canna-milk

- ¾ cup breadcrumbs

- 5 mozzarella cheese string sticks

- 1 egg

- 5 egg roll wrappers

- Oil, to fry, as required

Directions:

1. Whisk together eggs and canna-milk in a bowl. Place breadcrumbs in a shallow bowl.

2. Place one egg roll wrapper on your countertop with the one of the corners towards you.

3. Brush water on the 2 opposite edges (far away from you). Lay a cheese string stick on the corner closer to you and roll the wrapper along with the cheese stick up to one-third of the wrapper. Now fold the left corner inwards and the right corner inwards, over the rolled cheese stick.

4. Continue rolling to reach the corner farthest from you. Press the edges to seal.

5. Repeat steps 2 – 4 and roll the remaining egg roll wrappers over the cheese sticks.

6. Dip the rolled mozzarella sticks in egg, one at a time. Shake to remove excess egg. Dredge in breadcrumb and place on a plate.

7. Place a deep fryer pan over medium flame. Pour enough oil such that it is about 2 inches in height from the bottom of the pan. When the oil is well heated but

not smoking, 375° F carefully drop a couple of breaded mozzarella sticks in the oil. Cook until golden brown.

8. Remove the mozzarella sticks with a slotted spoon and set aside on a plate lined with paper towels.

9. Fry the remaining mozzarella sticks in batches.

10. Serve with a dip of your choice.

Weed Potato Chips

Preparation time: 5 minutes

Cooking time: 15 minutes

Makes: 2 servings

Ingredients:

- ¼ cup canna-oil

- 1 tablespoon Kernels popcorn seasoning of your choice

- 1 large potato, peeled, cut into thin slices

- 1 tablespoon salt to be used if you are not using Kernels popcorn seasoning

Directions:

1. Prepare a large baking sheet by lining it with parchment paper.

2. Place the potato slices on the baking sheet in a single layer.

3. Brush canna-oil over each of the chips, on both the sides.

4. Place the baking sheet in an oven that has been preheated to 400° F or until crisp and brown.

5. Take out the baking sheet, form the oven, and let it cool for 5 minutes. Sprinkle Kernels popcorn seasoning over the chips.

6. Serve.

Weed Biscuits

Preparation time: 15 minutes

Cooking time: 12 – 15 minutes

Makes: 20 – 25 servings

Ingredients:

- 1 ½ cups boiled, mashed sweet potatoes

- 3 cups all-purpose flour

- 2 tablespoons baking powder

- ¾ cup canna-butter, unsalted, cold, cut into small cubes

- 2/3 - 1 cup milk

- 4 tablespoons sugar

- 2 teaspoons salt

Directions:

1. Grease a baking sheet by spraying it with cooking spray.

2. Add about 2/3-cup milk and sweet potatoes into a bowl and whisk well.

3. Combine flour, baking powder, sugar, and salt in a bowl. Add butter and cut it into the flour mixture using a fork or a pastry cutter until crumbly in texture.

4. Add the sweet potato mixture and fold. If the dough is very dry, then add some more milk, 1 tablespoon at a time, and mix well each time using your hands

5. Form into smooth dough so add more milk if required.

6. Dust your countertop with some flour. Place the dough on it and roll the dough with a rolling pin. Cut into biscuits using a biscuit cutter.

7. Collect the scrap dough and form it into a ball. Re-roll the scrap dough and cut out some more biscuits. Keep doing this step until there is no more dough left to make biscuits.

8. Gently lift the biscuits and place on the prepared baking sheet in a single layer. Leave at least 1-inch gap between the biscuits. Bake in batches if required.

9. Place the baking sheet in an oven that has been preheated to 425° F for 12 to 15 minutes or until slightly hard and golden brown.

10. Remove the baking sheet from the oven and cool the biscuits completely.

11. You can brush with a little canna-honey or canna-butter on top if desired and serve.

Super Lemon Haze Mexican Guacamole

Preparation time: 20 minutes

Cooking time: 0 minutes

Makes: 8 - 10 servings

Ingredients:

- 8 Hass avocadoes, peeled, pitted, mashed

- 4 small heads garlic, peeled, minced

- Juice of 2 limes

- 2 teaspoons paprika

- 1 teaspoon cayenne pepper

- 2 sweet white onions, chopped into ¼ inch cubes

- 2 cups chopped small cherry tomatoes (¼ inch cubes)

- 2 ounces canna-olive oil

- 2 teaspoons chili powder or to taste

- Cracked pepper to taste

- Sea salt to taste

Directions:

1. Add avocado, garlic, lime juice, salt, spices, onion, tomatoes, and canna-olive oil into a bowl. Stir until well combined.

2. Cover and chill for a while for the flavors to set in.

3. Serve with vegetable sticks or crackers. This can be served chilled or at room temperature.

Pepper Crackers

Preparation time: 15 – 20 minutes

Cooking time: 10 minutes

Makes: 20 bite-size crackers

Ingredients:

- ½ cup flour + extra to dust
- 2 teaspoons fresh, chopped rosemary
- 1 ½ tablespoons olive oil
- ½ teaspoon baking powder
- 2 tablespoons ice water
- ½ teaspoon canna-olive oil
- 1/8 teaspoon salt or to taste
- Pepper to taste

Directions:

1. Combine salt, pepper, flour, and baking powder in a bowl.

2. Add rosemary and stir.

3. Combine ice water, canna-oil, and olive oil in another bowl and whisk well.

4. Pour into the bowl of flour and mix until dough is formed.

5. Knead into smooth dough using your hands. If the dough is hard and difficult to handle, cover the dough with a moist cloth for 5 minutes.

6. Knead until you get smooth dough.

7. Cover and set aside for 1 – 8 hours.

8. Dust your cutting board with some flour. Turn the dough onto the cutting board and roll the dough until very thin. Cut into 20 equal squares.

9. Line a baking sheet with parchment paper. Place the crackers on it in a single layer.

10. Place the baking sheet in an oven that has been preheated to 365° F and bake for 8 to 10 minutes or until light brown.

11. Remove the baking sheet from the oven and cool completely. On cooling, they turn crisp.

12. Serve with a dip of your choice.

Chocolate Crackers

Preparation time: 15 – 20 minutes

Cooking time: 10 minutes

Makes: 20 bite-size crackers

Ingredients:

- ½ cup flour + extra to dust
- 2 teaspoons fresh, chopped rosemary
- 1 ½ tablespoons olive oil
- ½ teaspoon baking powder
- ½ tablespoon sugar
- 2 tablespoons ice water
- ½ teaspoon canna-olive oil
- 1 tablespoon cocoa powder

Directions:

1. Combine sugar, cocoa powder, flour, and baking powder in a bowl.
2. Add rosemary and stir.
3. Combine ice water, canna-oil, and olive oil in another bowl and whisk well.
4. Pour into the bowl of flour and mix until dough is formed.
5. Knead into smooth dough using your hands. If the dough is hard and difficult to handle, cover the dough with a moist cloth for 5 minutes.
6. Knead until you get smooth dough.
7. Cover and set aside for 1 – 8 hours.
8. Dust your cutting board with some flour. Turn the dough onto the cutting board and roll the dough until very thin. Cut into 20 equal squares.

9. Line a baking sheet with parchment paper. Place the crackers on it in a single layer.

10. Place the baking sheet in an oven that has been preheated to 365° F and bake for 8 to 10 minutes or until light brown.

11. Remove the baking sheet from the oven and cool completely. On cooling, they turn crisp.

Lemon Crackers

Preparation time: 15 – 20 minutes

Cooking time: 10 minutes

Makes: 20 bite-size crackers

Ingredients:

- ½ cup flour + extra to dust
- ½ teaspoon grated lemon zest
- 1 ½ tablespoons olive oil
- ½ teaspoon baking powder
- 2 tablespoons ice water
- ½ teaspoon canna-olive oil
- 1/8 teaspoon salt or to taste
- 1/8 teaspoon red pepper flakes

Directions:

1. Combine salt, red pepper flakes, flour, and baking powder in a bowl.
2. Add lemon zest and stir.
3. Combine ice water, canna-oil, and olive oil in another bowl and whisk well.
4. Pour into the bowl of flour and mix until dough is formed.
5. Knead into smooth dough using your hands. If the dough is hard and difficult to handle, cover the dough with a moist cloth for 5 minutes.
6. Knead until you get smooth dough.
7. Cover and set aside for 1 – 8 hours.
8. Dust your cutting board with some flour. Turn the dough onto the cutting board and roll the dough until very thin. Cut into 20 equal squares.

9. Line a baking sheet with parchment paper. Place the crackers on it in a single layer.

10. Place the baking sheet in an oven that has been preheated to 365° F and bake for 8 to 10 minutes or until light brown.

11. Remove the baking sheet from the oven and cool completely. On cooling, they turn crisp.

12. Serve with a dip of your choice.

Sweet, Spicy, and Sativa Mixed Nuts

Preparation time: 5 minutes

Cooking time: 7 minutes

Makes: 4 servings of ¼ cup

Ingredients:

- 1 ¼ tablespoons canna-butter or canna-oil

- ½ teaspoon salt

- 1/8 teaspoon chili powder

- ¼ teaspoon dried rosemary, crumbled

- 1 ½ tablespoons dark brown sugar

- 1/8 teaspoon ground cinnamon

- A pinch cayenne pepper

- 1 cup nuts of your choice, toasted

Directions:

1. Prepare a baking sheet by lining it with parchment paper.

2. Add canna-butter into a skillet. Place the skillet over medium flame. When it melts, add salt, chili powder, rosemary, sugar, cinnamon, and cayenne pepper. Stir constantly until sugar melts.

3. Lower the flame and add nuts. Keep stirring until nuts are well coated with the mixture.

4. Transfer the nut mixture onto the baking sheet. Spread it evenly. Cool completely.

5. Transfer into an airtight container and store at room temperature. It can last for a week.

Cannabis Granola Bars

Preparation time: 25 minutes

Cooking time: 30 minutes

Makes: 8 – 10 servings

Ingredients:

- 3 tablespoons canna-butter
- 1 tablespoon maple syrup
- ½ teaspoon ginger powder
- ¼ cup raisins
- 1/8 cup hazelnuts
- ½ cup bittersweet chocolate
- 3 tablespoons honey
- 1 tablespoon brown sugar
- ½ teaspoon cinnamon powder
- ¼ cup chopped almonds
- 1 ½ cups rolled oats

Directions:

1. Take 2 pots of nearly the same (but not same) sizes such that the smaller one fits inside, the larger pot., The smaller pot should not touch the bottom of the bigger pot. It should fit well inside it.

2. Pour enough water into the larger pot such that it is 1/3 full. The water should not touch the smaller pot. Place the bigger bowl over medium flame. Let the water come to a boil.

3. Add chocolate into the smaller pot. Place the smaller pot inside the bigger pot.

4. Lower heat to low heat and let the water simmer. Once chocolate melts, remove the smaller pot from the double boiler.

5. In the meantime, add canna-butter, honey, cinnamon, maple syrup, sugar, and ginger powder into a small saucepan. Place the saucepan over low flame. Stir frequently until sugar dissolves completely.

6. Turn off the heat. Add raisins, oats, almonds, and hazelnuts and stir until well coated.

7. Prepare a baking sheet by lining it with parchment paper.

8. Spread the mixture on the baking sheet and press it evenly.

9. Place the baking sheet in an oven that has been preheated to 350° F and bake for 25 minutes.

10. Remove the baking sheet from the oven and cool completely.

11. Cut into bars of size 1 ½ x 4 inches. Drizzle the melted chocolate over the granola bars.

Hash Yogurt

Preparation time: 5 minutes

Cooking time: 5 minutes

Makes: 4 servings

Ingredients:

- 2 tubs flavored yogurt of your choice

- Butter or coconut oil, as required

- 0.03 ounce hash, crumbled

Directions:

1. Add butter into a pan. Place the pan over low flame. When butter melts, add hash and stir. Cook for a couple of minutes, stirring often. Turn off the heat.

2. Divide into the tubs of yogurt and stir.

3. Serve.

Hush Puppies

Preparation time: 10 minutes

Cooking time: 20 minutes

Makes: servings

Ingredients:

- 2 cups canna-milk

- 3 cups self-rising cornmeal

- 1 cup self-rising flour

- 1 teaspoon baking soda

- 1 teaspoon salt

- 2 eggs

- Vegetable oil for deep frying, as required

Directions:

1. To mix dry ingredients: Add cornmeal, flour, baking soda, and salt into a mixing bowl and stir.

2. To mix wet ingredients: Add canna-milk and eggs into a bowl and whisk well.

3. Combine the wet and dry ingredients in a bowl. Mix well to form a thick batter. Set aside for 10 minutes.

4. Place a small deep pan over medium heat. Add enough vegetable oil to cover at least 2 inches in height from the bottom of the pan.

5. When the oil is hot, about 370° F (it should not smoke), drop the teaspoonful's of batter in the hot oil. Add as many as can fit in the pan.

6. Fry until golden brown all over.

7. Remove with a slotted spoon and place on paper towels.

8. Fry the remaining in batches in a similar manner.

9. Serve with ketchup or any salsa.

Marijuana Hot Wings

Preparation time: 15 – 20 minutes

Cooking time: 15 – 20 minutes

Makes: 4 servings

Ingredients:

- 1 pound fresh chicken wings, discard wing tips, cut into wingettes and drummets

- ¼ cup red hot sauce

- ¼ cup canna-butter, melted

- Vegetable oil for deep frying, as required

- Ranch dressing to serve

Directions:

1. Pat the chicken with paper towels to dry.

2. Place a small deep pan over medium heat. Add enough vegetable oil to cover at least 2 inches in height from the bottom of the pan. Let the oil heat.

3. In the meantime, combine canna-butter and red-hot sauce in a shallow bowl.

4. When the oil is heated to 375° F (it should not smoke), add chicken wings in the pan of hot oil. Add as many as can fit in the pan.

5. Fry until golden brown all over.

6. Remove with a slotted spoon and place on paper towels.

7. Fry the remaining in batches in a similar manner.

8. Add the chicken wings into the bowl of sauce mixture. Toss well.

9. Serve with ranch dressing.

Peanut Butter Balls

Preparation time: 15 minutes

Cooking time: 20 minutes

Makes: 30 – 40 servings

Ingredients:

- ¾ cup peanut butter

- 2 cups confectioner's sugar

- 1 cup semi-sweet chocolate chips

- ½ cup canna-butter, chilled

- 2/3 cup Graham cracker crumbs

- ½ tablespoon shortening

Directions:

1. To melt chocolate: Take 2 pots of nearly the same (but not same) sizes such that the smaller one fits inside, the larger pot., The smaller pot should not touch the bottom of the bigger pot. It should fit well inside it.

2. Pour enough water into the larger pot such that it is 1/3 full. The water should not touch the smaller pot. Place the bigger bowl over medium flame. Let the water come to a boil.

3. Add chocolate and shortening into the smaller pot. Place the smaller pot inside the bigger pot. Stir occasionally.

4. Lower heat to low heat and let the water simmer. Once chocolate melts, remove the smaller pot from the double boiler.

5. To make peanut butter balls: In the meantime, add canna-butter and peanut butter into a mixing bowl. Mix with an electric hand mixer until well combined.

6. Mix in the Graham cracker crumbs. Make small balls of the mixture of about 1 inch in diameter and place on a baking sheet lined with wax paper.

7. Dip the peanut butter balls in the melted chocolate, one at a time. You can use a toothpick to do so. Place the dipped ball back on the baking sheet.

8. Do this with all the balls. Place the baking sheet in the freezer and freeze until the chocolate sets.

9. Transfer into an airtight container and refrigerate until use.

CONCLUSION

As you just witnessed, there are several different and delicious ways to incorporate Cannabis for medical uses in your food. Stronger doses of cannabis are definitely not recommended. To be safe, go in for smaller doses.

The recipes given in the book are meant to be taken in small doses. You can increase or decrease the dose depending on your personal needs, but beware of the side effects that might ensue with increased dosages. We hope these recipes help you manage your medical problems in a scrumptious manner.